STOLEN FOR PROFIT

STOLEN
FOR PROFIT

How the Medical Establishment
Is Funding a National
Pet-Theft Conspiracy

by JUDITH REITMAN
Introduction by Dr. Michael W. Fox

PHAROS BOOKS
A SCRIPPS HOWARD COMPANY
NEW YORK

Library of Congress Cataloging-in-Publication Data

Reitman, Judith.
    Stolen for profit : how the establishment is funding a
national pet-theft conspiracy / by Judith Reitman.
        p.   cm.
    ISBN 0-88687-676-1 : $19.95
    1. Animal experimentation—United States.   2. Pet
theft—United States.   3. Laboratory animals—United
States.   I. Title.
HV4931.R45   1993
364.1'62—dc20   92-38652   CIP

Printed in the United States of America

Pharos Books
A Scripps Howard Company
200 Park Avenue
New York, N.Y. 10166

10   9   8   7   6   5   4   3   2   1

Pharos Books are available at special discounts on bulk
purchases for sales promotions, premiums, fundraising or
educational use. For details, contact the Special Sales
Department, Pharos Books, 200 Park Avenue, New York,
NY 10166.

For the victims

CONTENTS

America's pets are disappearing. They are vanishing from neighborhood streets, yards, and kennels in every town and city in this country in numbers that are startling. Over 10,000 dogs were reported missing within a six-month period in Rochester, New York; 700 dogs vanished within an eleven-month period in the Orlando, Florida, area; 500 dogs and cats disappeared from Honesdale, Pennsylvania, in six months; 1000 dogs were missing from Indianapolis in one month. Hundreds of thousands more dogs and cats are "adopted" through newspaper ads that their owners place—"Free to a Good Home" ads—but their destinations are not homes.

Each year as many as two million family pets are stolen and sold into a gulag of a nightmare that trades them into a black market for pet stores, puppy mills, dog-fighting rings, and satanic cults. But by far the most valued and reliable buyer for these animals is the medical research industry, which can pay premium tax dollars for preferred laboratory subjects: family pets, no questions asked.

These "companion animals" will be burned in radiation tests to study the effects of nuclear fall-out. They will be shot at close range in studies of head trauma injury. They will be addicted to drugs, nicotine, and chemical solvents. They will be force-fed toxic household products for liability insurance purposes, and they will be dissected in classroom teaching exercises.

But the issue here is not the morality or efficacy of biomedical research, nor is it animal rights, though both are certainly important corollaries. The issue is crime. Institutionalized crime. There are laws in place to protect people's pets, but they are not being enforced because government is in a long-standing economic relationship with an industry that depends on the delivery of stolen merchandise. That industry is a segment of the multibillion dollar medical research industry. The delivery systems are dog dealerships licensed by the United States Department of Agriculture. And the stolen merchandise is people's pets.

High-demand animal research has created a multimillion-dollar market for dogs and cats that is run by syndicates as hierarchical and well protected as the Mafia. Nowhere is the scope of this network more visible than at dog auctions in the Midwest, where dealer and buncher trucks from virtually every state in the Union converge to swap stolen dogs and cats, switch license plates and cab loads at state borders, phoney-up paperwork, and fill orders from nearly all the universities, hospitals, products-testing companies, even military bases in the country that conduct research using dogs and cats.

For decades, the power and prestige of the medical research industry have ensured animal racketeers hefty incomes. Local and federal agencies have ensured these criminals impunity. Together, government and industry have sabotaged investigations of dog dealers, used intimidation lawsuits to silence victims, and threatened the constitutional rights of outraged citizens.

Stolen for Profit is the story of the courageous men and women who are risking their lives battling the pet-theft syndicates.

Ultimately, this is a story about how the medical industry and the government have betrayed us.

•   •   •

My personal journey into the nefarious underworld of pet theft began in the summer of 1990. While reporting an article on stolen pets for a national magazine, I attended a dog auction in backwoods Missouri. Six hours after leaving St. Louis, where my plane had landed, my "guide," a local auction watcher, and I pulled up to the mud-gouged parking lot of what appeared to be a county flea market. During that pre-dawn trip I had been given my first lesson in the terminology of pet theft: bunchers, "junk dogs," bait, "serum trucks." I had also been given a warning: "Don't look any of the dogs in the eye."

By the time we left the auction, a half-dead coon hound pup purchased from a "bait bin" was sprawled on my lap, and I knew my life would never be the same.

I had been dressed for auction weekend like the rest of the locals: worn, dirty jeans, feed cap. But in the pocket of my T-shirt, camouflaged by a crumpled packet of cigarettes, was a miniature tape recorder. It was too early to be fully awake, and the auction site, hidden in a clearing in the woods, had a nightmarish quality. The smell of feces, sweat, and turkey "nuts" cooking at camp fires hung in the air. At least one hundred dogs looking shell shocked were attached chain-gang style to the fenders of pick-ups and U-Hauls slung low in the fetid mud amid rotted lean-to's, where AK-47s and other semi-automatic weapons were sold to anyone with the cash.

"We got dogs from nearly every state in the Union," the auction owner told me. They were brought here for sale to "serum trucks" for use in medical research; the suppliers, licensed by the U.S. Department of Agriculture.

In the months to come I would learn that there was an even more horrifying behind-the-scenes reality that was sending stolen pets into laboratories. A remarkable woman, then 80 years old, gave me my first lesson in the politics of dealing stolen pets. "Organized crime sanctioned by the U.S. government," was how Mary Warner termed pet theft for sale to research. She is known as the "godmother" of the front line fighters, the woman who leads the struggle against the criminal network known as the "Dog Mafia," and the founder of Action 81, a grass-roots organization that tracked pet-theft networks nationwide.

This book contains hundreds of interviews I conducted during the course of nearly three years with front line fighters; local, state, and federal law enforcers; auction owners; dealers and bunchers; government officials, including those from USDA, the Animal Plant Health Inspection Services, and Regulatory Enforcement and Animal Care and its Task Force; medical researchers; reporters; and scores of victims of pet theft in virtually every state. The process of writing about pet theft was similar to writing about espionage or organized crime. It involved death threats, calls placed to me from phone booths at all hours of the night, internal memos sent with no return address, "coded" messages, the hushed, frightened voices of many insiders who felt a moral duty to speak out against a system that in their eyes defiled the democratic process. Those sources agreed to speak if guaranteed anonymity; they feared for their lives and careers.

The years spent on *Stolen for Profit* have been enlightening, and emotionally wrenching. If readers can be spared the tragic theft of their pet, then the book's mission will have been fulfilled.

Judith Reitman

ACKNOWLEDGMENTS

This book owes its existence to Matthew Bialer of the William Morris Agency who saw the potential of the material, found a courageous publisher, and through friendship and diplomacy nurtured this project to its completion. I am forever indebted to my family for their loving support, with special gratitude to my father for his generosity of spirit. I am deeply grateful to Christine Henkel for her brilliant perspective over many lifetimes, to Christine Bounce for her devotion and tireless commitment to a Cause, to Bill Cotreau for his wry wisdom. I want to thank my editor Eileen Schlesinger for her resiliency and receptivity during the many upheavals that accompanied this project, and Veronica Johnson for her thorough and insightful copyediting.

Stolen for Profit has been a collective effort on the part of many individuals and I am grateful to those who assisted me along this journey: the victims of the Dog Mafia and the front line fighters who opened their homes and hearts to me; Susan Chasworth, Los Angeles County District Attorney's office and Norm Wegener, LA City Attorney's office; Suzanne Roy; Deputy Sheriff Wayne Tyler; Jim Flasch and Mel Hanks; Bob Baker; Christine Stevens; Dr. Jan Moor-Jankowski; and Peter Bloch, for the courage to print the "Dog Mafia" article in *Penthouse* that began it all. I greatly benefited from the resources made available to me by the Freedom of Information Office, the Physicians Committee for Responsible Medicine, the Medical Research Modernization Committee, Protect Our Pets, COMBAT,

National Institutes of Health, OMB Watch, Reporters Committee for Freedom of the Press, and United Action for Animals.

Thank you to Matthew Weber for his legal expertise and to Christopher Schelling for his editorial perspective at the proposal stage of this book.

For their enduring friendship I want to thank Jeanne McDowell, Linda Bruner and Gene Ruscigno, Bob and Laurie Schrieber, Neal Smith, Jim Motovalli, Jim Mason, Alexandra Randall, Ron Kerner, the Raithels, the Aiellos, and Philippa and Ray Hill.

For the lessons in faith I am grateful to Carmen Santiago. For the lessons in courage, Mary Warner.

Introduction

by Dr. Michael W. Fox

After reading this book, I felt violated and angry—and a sense of moral outrage to which I am no stranger when it comes to how some people treat dogs and other animals in society today.

I was shocked into disbelief by the factual details and purported action of the evil characters in this book.

It is a spellbinding account, and as the plot is unraveled there is revealed a heinous and fraudulent scam: pet theft for profit. It is a practice by which criminals deceptively pose as caring and loving prospective adopters for thousands of beloved companion cats and dogs who, for various reasons, have to be separated from their original human custodians.

I have debated in public and on national television with people who say, "Why not let the biomedical industry use all those millions of lost and abandoned cats and dogs that are destroyed every year in animal shelters across the country?" After all, these people say, those animals were going to die anyway.

Many shelters across the country are still open to researchers wanting dogs and cats. And in those areas allowing "pound seizure," pet theft is still high, many times even higher than in areas where pounds are closed to researchers, as *Stolen For Profit* reveals. And when, in

Dr. Fox is the Director of the Center for Respect of Life and Environment. He is the author of over forty books on animal behavior and concepts including *Inhumane Society: The American Way of Exploiting Animals* (St. Martin's Press, 1990).

the late 1960s, most shelters were open to researchers, two million pets were still stolen. This kind of availability of companion animals perpetuates much research using animals, research which is of questionable value, often undertaken only for the grant money; the use of dogs and cats is controversial not only from a moral but scientific perspective.

Many sophisticated alternatives to live animals exist for researchers, as well as for students wanting to dissect. There are models and films so detailed that many medical schools have substituted them for the so-called "dog labs" as being more appropriate alternatives for the use of animals. A side effect of utilizing such alternatives is that the medical students are learning a greater respect and compassion for all life forms.

The government, the biomedical research industry, and the criminals that comprise the "Dog Mafia" have a common attitude toward nonhuman life. They share the view that nothing in animal form is sacred, that just as the human/companion animal bond is not sacrosanct, neither is our collective responsibility toward the rest of the animal kingdom. For these people, money and power reign supreme.

These sentiments allow the Dog Mafia to thrive and the government and biomedical research establishment to continue to condone the use of animals for the most trivial and often painful experiments and classroom exercises.

Judith Reitman mentions two other atrocities that involve the exploitation of dogs—namely, illegal dog fighting and the legal "puppy mill" breeding farms for the commercial mass production of purebred dogs for the pet trade. Both horrors profit from the pet-theft trade. In my mind, the fact that these atrocities, like pet theft for profit, are so widespread and institutionalized is indicative of the dysfunctional nature of society today.

Clearly, we must change society if we are ever to stem the rising tide of inhumanity that profits by bringing so much evil and suffering into the world. The profit motive is strong; so strong that the anticipated costs to the biomedical industry of having to comply with recent amendments to the Animal Welfare Act resulted in the government's delaying their implementation. These amendments would provide laboratory-caged dogs with regular exercise and primates with conditions that better satisfy their behavioral and psychological needs. The delaying tactics were initiated, in part, by the President's Council on

Competitiveness, which is now on public record as placing the interests of industry over environmental, social, and animal-protection concerns. Inhumanity is evidently not limited to one region or segment of society. It is a widespread problem within the very fabric of society today and in the highest circles of government itself.

I hope that this book will be read by many and will lead to more than "tighter" regulations over the use of animals by the biomedical research industry. As a society we need a change of heart, for surely no good ends can ever come from evil means; after all, there is no greater authority than the heart.

ONE

Crimes Against the People

"This is a guerilla war like the one fought in Vietnam. A government is at war against the People and the People are fighting back. In every village and town and city, people are tired of being lied to, and they are fighting back."

—*Mary Warner*
Founder, Action 81

August 9, 1991
San Fernando Courthouse, County of Los Angeles
The call from the court clerk came into the District Attorney's office at 2:48 p.m. Within moments, half a dozen phone lines in the third floor office of the San Fernando Courthouse were jammed with press calls, and the interoffice speaker was blaring, "Susan Chasworth, come to the front desk immediately."

During the two and a half days the jury was out deliberating, the DA's staff had been lulled back into its normal routine: felony, robbery, murder, and rape cases that occurred within the 359-mile radius of the San Fernando Valley. In 1989, when the Ruggiero case was filed as a felony, the county was prosecuting 72,981 felonies and 273,303 misdemeanors. This year, 1991, the DA's caseload was even heavier.

Despite the ever-increasing crime rate, the county court had never encountered the kind of crimes now being adjudicated before Judge David Schacter.

By 2:50 on this unusually cool summer afternoon, the prosecutor's office was transformed into the nerve center of a case which defied credibility. This was the San Fernando Valley of Los Angeles, not backwoods Missouri. This was the region that matched the Spaniards' vision of heaven on their so-called "Mission to Paradise" in 1769. Now, centuries later there were a dozen, upscale malls in this modern-day paradise for the comfort and convenience of 1.3 million residents; one mall for nearly each of the twelve LA bedroom communities. TV and movie stars built palatial mansions in the panoramic Hollywood Hills. "California

5

Mothers of the Year" chauffeured their kids from award-winning Valley schools to music and dance lessons, and families spent weekends swimming at Malibu, skiing at Wrightwood, picnicking at state parks.

No one would have guessed from looking at the three well-dressed defendants that they were part of a network that ran deep into the heartland of America, a network rooted in the bloodied mud of Midwest dog auctions. The People's case against these three had to build carefully because an entire system was at stake, a system that was sanctioned by the United States government.

At 2:55, a crowd converged on Superior Court Department A: the defendants and their court-appointed attorneys, press and camera crews with bulky lighting props, and city and county prosecutors who left their own offices to hear the verdict. Many of the victims had been alerted, and several who lived nearby rushed to the courthouse. There were also curious onlookers who had been following the highly publicized case.

The revelations in the courtroom had provoked outrage and shock. The sickening crimes struck a primal nerve that galvanized the entire Valley, and the communities had fought back.

"Susan Chasworth, please come to the front desk," the loudspeaker in the DA's office blared; the tone was urgent.

At 3:05 the defendants were taking their seats at the defense table. Each was flanked by his own attorney. Although the lawyers had their individual agendas, they shared a common goal: They would demonstrate that their clients were, themselves, victims.

"Three-thirty on a Friday, as predicted," Eli Guana remarked to his client, Frederick "Rick" Spero. Rick, age forty-six, a former real estate broker, saw little irony in the timing. His usually swarthy complexion was ashen. The shadows around his brown eyes had darkened over the months.

Rick looked at Barbara Ruggiero, hoping for some sign of comfort, but her face displayed no emotion. His former fiancée remained an enigma to him.

He pressed his hands together and closed his eyes. He seemed to be praying.

On the opposite side of the defense table, Ralf Jacobsen, twenty-eight, was giving instructions to his young German wife, who sat behind him in the gallery. The girl wore a black mini dress and a tight tangerine sweater. She was crying as her new husband spoke in rapid German, his second language.

Ralf Jacobsen, AKA Steve Jacobs, Mike Johnson, Steve Johnson, Mike Rogers, Paul Bruttel, Craig, Alexander Latham, had hoped to become an attorney. His victims described him as the "blond, crew-cut, All American type," but Ralf's once rosy complexion had since taken on a gray cast. Already slight of build, he seemed to have diminished in size since the trial began.

Ralf did not speak with, or look at, his co-conspirators. Among them, sitting to his left, was the woman he and Rick Spero had loved.

Barbara Ann Ruggiero, twenty-eight, AKA Kimberly Lynn Christianson, appeared composed and confident. Her attorney, Lewis Watnick, had argued that she was the victim of mob violence, but it was difficult for most of the onlookers to picture Barbara as a victim. There was something imperious about her bearing, something chilling about her fixed smile.

Valedictorian of her high school, Barbara's ambition was, she had said, "to be around happy animals." Yet she had mobilized her coterie of lovers to steal hundreds of family pets for sale to medical research. The scheme netted her $17,500 in one month alone.

As Barbara smiled at her attorney, her cheeks dimpled. She wore no makeup and looked very demure.

"How does that girl get so many men to do those things for her?" a woman in the gallery wondered aloud. There had been much speculation about Barbara's method of coercion.

Barbara's sister, Joanne Fenn, sat in the gallery. Fenn, thirty-six, a heavy-set woman with frosted hair and garishly painted nails, stared icily at the onlookers.

"Susan Chasworth, please come to the front desk," the speaker in the DA's office again blared. "You have a verdict."

The Deputy District Attorney was racing to Department A. She had attended an impromptu conference at an adjoining building, but even at that meeting she had been preoccupied with this case.

Waiting for a verdict was tough, very tough. Snapshots of the witnesses and the defendants and snatches of testimony clicked repeatedly in her mind. It had been a long haul, but Susan had been prepared to go the distance. She had expected the long hours, looked forward, in fact, to the voluminous documentation. She had easily endured the marathon trial; it was no different from the marathons she ran and biked.

Susan, or "Flash," was a beautiful woman with waist-length strawberry-blond hair, blue eyes, and delicate features. New York born and raised, the thirty-three-year-old prosecutor was, by nature, serious and analytical. She had been attracted to the legal profession because of its logic and clarity. In the Ruggiero case, she was especially grateful for the judicial forum as a way of containing her own turbulent feelings. She found the crimes morally and emotionally repugnant. The motive was simple: greed. Trying to get something for nothing. These people did not care about the damage they had done to hundreds of lives.

"I hope I don't get up on the stand and start bawling," Norman Flint, one of the victims, had confided to her. Flint was well over six feet, a 250-pound construction worker. It had been painful to watch him fight back tears on the stand. He had told her, in private, that he was prepared to go to jail for killing Ralf Jacobsen. "But that wouldn't have brought Bear back, so I told him I was going to beat the shit out of him in court."

Susan understood what Flint meant.

The corridors were virtually empty when Susan reached the mezzanine. It had been a media circus earlier in the day when Rick James was brought to trial from county jail. The forty-three-year-old "King of Funk" had allegedly tortured a young woman with a hot cocaine pipe and forced her to have sex with his twenty-one-year-old girlfriend. That case drew the press like flies.

People liked sensationalism and Susan had had ample opportunity to go for high drama in the Ruggiero trial. She could have brought the collars into court and piled them high on the clerk's desk. The red and yellow and black collars with their bronze-plated tags from the United States Department of Agriculture, the government agency that had licensed the defendants. The small collars, the ones that belonged to the cats, were most pathetic.

She could have stacked the clerk's desk with 140 photographs. One in particular made her cry, Paul Iverson's young German Shepherd, PJ, sitting in his living room, trying to be good. The dog looked so little, and the picture was so big.

She could have brought into testimony Barbara's electric cattle prods, the frozen puppies that were fed to the pet python, and the cats that were microwaved into feed.

She could even have brought into court one of the few survivors. She could have brought in Wiggles.

But Susan had decided against the collars and the photos and the shock tactics.

Frankly, they were not relevant. Sensational, yes, but she had no intention of sinking to the level of taking cheap shots at the defendants. She was going to try a clean case. The emotions were implicit. So were the lies.

"Have you ever used a false name to obtain or sell animals?"

"Never."

"Does the name Kimberly Lynn Christianson mean anything to you?"

There had been seventy People's exhibits, each exhibit with its multiple parts. Evidence and heart-wrenching testimony told a story about crimes so deliberate and coldhearted they defied comprehension. Defied comprehension, if you were a human being with the capacity to feel guilt and shame.

And there it was—what it came down to, whatever the nature of the crime: conscience.

Susan thought that Barbara Ruggiero, who looked so innocent, did not consider, was perhaps incapable of considering, the effect her actions had on other people's lives. Barbara understood only the consequences to herself.

At 3:16, Susan swung open the courtroom doors. What she had come to call "the most exciting moment in human life" was just moments away. The People of the State of California versus Barbara Ann Ruggiero, Frederick John Spero, and Ralf Jacobsen was now in the domain of forces over which she had no control—the unfathomable human mind and the capricious human heart.

"Call in the jury," Judge David Schacter instructed the bailiff.

Regardless of the type of case being tried, juries are invariably difficult to read. Susan turned, instead, to look at Barbara. She thought about the many lives this woman had damaged, and she knew that Barbara felt no remorse, no guilt at all. Barbara's face displayed nothing—no fear, no expectation.

There were hundreds of thousands of ordinary citizens across the country who were victims of a network that allowed people like Barbara Ruggiero to profit handsomely. But Barbara and her cohorts had not counted on another network forged by a woman who now waited anxiously by her phone, three thousand miles away.

August 9
Northern Virginia

In a small outbuilding nestled in the emerald-green hills of Virginia horse country, Mary Warner waited for a phone call from the San Fernando Courthouse. The interior of her office was chaotic with documents and photos accumulated over the course of 30 years. As Mary waited, she played back dozens of phone messages.

"We figure it's a ring of about fifteen people in Kentucky and Indiana, Mary." Mary recognized Sue K.'s voice. "They're packin' Huskies, Shepherds, 150 big dogs per trip in three pickups, and bringin' them to the South Carolina flatlands." A ring of poachers was stealing dogs to stake out for alligator bait in the swamps. "They promise to shoot and burn out people if they rat on them."

Patty in Texas called: "We found the source of the bait for the dog-fighting rings, Mary. A buncher consortium. Dope and dogs."

But the bulk of messages that flooded Mary's answering machine were from front-line fighters tracking the dealer/research circuit for Action 81, Mary's pet-theft tracking network. "Mary," Linda Elliot drawled. "I got Deputy Sheriff Tyler over to that buncher camp. The place stinks. I think they're eatin' what they ain't sellin'." Mary smiled as she thought of Linda, a tough gal who looked like a beauty queen. Linda had singlehandedly busted the buncher encampment of an Arkansas kingpin.

Dwayne Reitz called to report his group's vigilant surveillance had paid off. The warehouses on the New York/Pennsylvania border were dealer waystations. "We were right. One of the drivers said they're supplying New York labs. A couple of people here found their missing dogs in the shed, but they're too afraid to prosecute. Can you blame them?" Dwayne and his family had been plagued by death threats. "These people don't fool around," he once told her.

Debby in Missouri left a message about "a run on big-chested dogs. We figure the call is out from cardiovascular researchers. We're getting two hundred calls a week for the past three weeks, people missing big dogs—Labs, Shepherds, Huskies. Animal Control's taking the dogs right out of people's yards. Have you got a contact in Arkansas for us?"

Pound managers were on the take, and with those kind of numbers, so was the local law.

The schemes and numbers did not shock Mary Warner. Few horror stories shocked her anymore. The ever-growing list of casualties tacked on her wall was a tragic reminder that nothing had really changed: over 10,000 dogs missing in Rochester, New York, within six months in 1983; seven hundred dogs missing in eleven months in Orlando, Florida, in 1985; 985 dogs and cats missing within eleven months in 1987 in Concord, North Carolina; over 1000 dogs and cats reported missing in Indianapolis in one month in 1989. In 1990, in Columbus, Georgia, 2500 dogs and cats had been reported missing: 5000 in two consecutive years.

Many of the owners of these pets had learned the painful truth about their destinations: Mayo Clinic, Harvard, Yale, University of California at Los Angeles, Washington University in St. Louis, University of Missouri, University of Virginia, University of Minnesota, University of Washington in Seattle, Gore Labs, Letterman Army Institute, University of California at Davis, 3M Corporation. . . .

The men and women who delivered these family pets into laboratories were licensed to do so by the United States Department of Agriculture. For as little as $40 plus a $10 application fee, virtually anyone could become a USDA-licensed purveyor of stolen pets for sale to medical research, and make a small fortune in the process. Fact was, as buncher Ralf Jacobsen had known, a license was not even really needed.

In Los Angeles, 140 pet owners had only recently discovered the horrifying reality that dogs and cats in laboratories are stolen pets. But why did it take such tragedy before the public, which pays for this research, was allowed to look behind the laboratory doors?

Mary had been asking that question for nearly thirty years.

The sun glittered gold and amber over the Shenandoah River as Mary jotted down the last of her messages. It was nearly 6:30 P.M.; 3:30 LA time. On this sultry summer evening, when the scent of newly mown grass rose from the moist fields and the honeysuckle hung heavy on the bushes, Mary was waiting for far more than a verdict in the Ruggiero case. She waited for a vindication of thousands of front-line fighters who were risking their lives battling the Dog Mafia.

Mary Warner would settle for nothing less than an indictment of the U.S. government on charges of conspiracy.

TWO

Free to a Good Home

"Barbara said that the animals were being sent to a person
who placed animals in the movies, as bomb sniffers, to
old folks homes to be with people, to farms as mousers."
—*Trial testimony of Terry Phillips*
caretaker, Budget Boarding Kennels

November 1987
City of Los Angeles

"Jesus Christ! Does he have to slam the damn cages?" Terry Phillips
grabbed the alarm clock beside his pillow. Five A.M. It was hard
enough falling asleep with seventy dogs barking and crying and fighting,
without the Mystery Man rattling around the groom room at dawn.

"Calm down, honey. Barbara told us——"

"Barbara told us a lot of things that don't make no sense," her
husband snapped. "For a lousy $150 a week we got to put up with
more shit that there is in the kennels. And this mama's boy coming
in at all hours of the night. Where are we going to put more dogs?
The kennels are jammed full."

The ruckus in the groom room woke the dogs who had been asleep
in their pens. They joined the uproar the new arrivals were making,
barking in the room off the kitchen.

There was no point in his trying to sleep.

Terry flung the covers from the makeshift bed of blankets and quilt
piled onto the living room shag carpet. "I'm going out for a walk,
baby," he told his wife.

"I'll come out with you," Cindi mumbled. But she turned on her
side and pulled the covers over her shoulder.

Terry tossed a handful of wood chips they had collected from the
neighboring dump site into the fire. The flames snatched the dry
kindling and flung a bright orange light onto the bedding, onto Cindi's
dark hair. Terry felt a stab of guilt for losing his temper.

They had been married only two years, and most of that time they had been weighed down by financial problems. And now Cindi was battling for custody of her son from a previous marriage when she was only seventeen. She was twenty-eight, three years his senior, and, Terry thought as he watched her sleeping, life was not getting any easier for his wife.

Money. If they weren't in such tight straits they would have left this dump weeks ago.

Terry slipped into his jeans and pulled a sweater over his tousled blond hair. He found his workboots near Monty's cage. Barbara's pet python slithered toward him, flicking its tongue against the glass walls of the aquarium.

Terry shuddered. The whole place gave him the creeps.

Terry quietly opened the bedroom door. Thai and Kimmo were sleeping peacefully on the bed. Barbara did not want them running loose, and he and Cindi were sure not going to put their dogs in those pens. "Kennels," Barbara called them. They were nothing more than oversized chicken coops and cement runs roofed with tin remnants or termite-ridden plywood. Rusted nails, broken pipes, and wire littered the kennel area.

Even after the runs were scrubbed down they smelled disgustingly of urine and loose feces. Poor dogs, Terry thought. No wonder they were always on edge. It was like living in a bathroom with four or five other people shitting and pissing on you.

The bedroom, which could be closed off, was the only solution for Thai and Kimmo. What the hell, let the dogs have the bed. Hopefully they would all be out of there soon.

The pack of Marlboro 100's in his jeans was empty, but he remembered there were some cigarettes in the kitchen. The fluorescent light above the countertop rat aquarium was on. As he rummaged through a drawer, Terry watched the rodents scurrying about. Those not destined for Monty's supper were sold to the movies, Barbara told him. He did not bother asking what her rodents did in the movies. The less he knew about the business of Budget Boarding, the better.

Budget Boarding Kennels was in Sun Valley, about a half-hour drive from downtown Los Angeles, up the Golden State Freeway and off the Antelope Valley Freeway, a southern California superhighway bordered by the San Bernadino Mountains and desert brush. A northeast suburb of the San Fernando Valley, Sun Valley was zoned for both residential and commercial property.

Budget's neighborhood was largely Hispanic, Oriental, and Black. Stucco houses, many sprayed with graffitti, tool factories, and meat-processing plants were crowded together along the narrow streets off Tuxford Boulevard. The area's lax noise regulations made it attractive to kennel owners who provided boarding and grooming services, and who also moonlighted as puppy-mill breeders. Along Norris, Bradley, and Astoria were dozens of such filthy backyard kennels bordered by chain-link and battered wood fences.

Also scattered along the industrial streets were animal welfare organizations which specialized in rescuing certain breeds and finding them homes. They, too, were overcrowded, but the animals appeared well cared for.

Budget Boarding at 8916 Bradley Avenue shared the corner lot with Von Konstanze Kennels. Both kennels were owned by Bridgette Jacobsen. Her son, Ralf, delivered animals to Budget at night because he did not want his mother to know he was working for Barbara. The German woman, who ran a Rottweiller import business at Von Konstanze Kennels, hated Barbara Ruggiero.

Long feuding rivals for the affection of young Ralf, the women also fought over the condition of Budget Boarding, which, despite Bridgette's constant haranguing of Barbara, was never improved. Barbara made no effort to clean up the broken glass and assorted rubbish in her kennels. To add to the chaos, a construction dump and quarry located at the rear of Budget did not appear to observe its boundaries. Budget's roofing looked as if it had been constructed by a strong wind that had flung rotted wood and nails over from the dump site.

Late in October, Terry had been thumbing through the unemployment ads section of the *Recycler*, a local newspaper. One caught his eye: "Salary and free room and board for work at a kennel." He had his GED high school equivalency and Cindi had quit school during junior high. They had no great aspirations, but they did love animals.

Barbara Ruggiero seemed a little harsh to him, but the job sounded easy. "All you have to do is clean the runs at seven in the morning, and feed and clean up the dogs," Barbara told them. Cindi would be in charge of the cat room. At night they'd have time to themselves.

The house that was attached to the kennel desperately needed cleaning, too, but it was larger than any they had lived in before.

That was only two weeks ago. Life was becoming more intolerable each day. The only consolation was working with the dogs and cats they had become so fond of. And it was a good cause, placing unwanted animals in nursing homes or in families with kids. And Barbara said the movie studios needed healthy, good-tempered animals.

When Terry walked into the kennels, the dogs immediately quieted. "So all you want is some company," he said. Some of the dogs softly whined. Terry could see their eyes glittering in the half light, their tails wagging. They were so hungry for affection.

The evening was unusually warm for October. A sliver of moon illuminated the telephone wires that criss-crossed the sky. Terry turned over a five-gallon plastic bucket they used for washing down the kennels and sat near Flicka's run. Flicka shared the space with four big dogs.

"Hi, honey. Hi, Flicka." Cindi had come out, wrapped in a blanket. Hearing her name, Flicka wagged her tail.

"She's so pretty," Cindi said. "Who would abandon her?"

Flicka was a very pretty dog. Black with white markings on her chest and tips of white on her paws, she looked like a Collie mixed with some Shepherd. Barbara said the lady who boarded Flicka never came for her and never paid her last bill.

"Wonder why Flicka's name is on that chalk board?" Cindi wondered aloud. It had been there since they arrived—FLICKA, spelled in big white letters on a board near the groom room.

"You ask her, baby."

"Yeah, right." Cindi snuffed out her cigarette. "How many animals in the groom room?"

"I didn't look. It's too early to get depressed." An orange cat scooted past them. "Charlie's on the prowl, Cin. Better check the bed before we get in."

Cindi grimaced. Charlie had a habit of leaving his mouse trophies in their bed. But he was so loving, and they had sort of adopted him.

"How do you think Barbara makes any money, Cin? She keeps getting more animals, but nobody's come to adopt them."

"There's that guy Stephens up north—in Oregon, I think she said. Isn't he the one who finds homes for them?"

"I thought Stephens was her lawyer. Anyway, if her boarders aren't paying, how is she going to keep this place going?"

"She inherited a lot of money after her mother died, remember?"

Barbara had told them her mother had been killed, murdered in late 1984. She had not given them any details, only that the perpetrator had not been found. It sounded gruesome, but the Phillipses had enough to do at the kennels without getting involved in Barbara's personal life.

Cindi recalled the day they moved in. She had been cleaning the stove, scrubbing off the fungus and grease, when what looked like a trap door snapped open and a wad of tightly wrapped bills popped out. There must have been hundreds of dollars. She put the bills in an envelope and left them on Barbara's desk.

Barbara told them she had a condo in Sylmar where she lived with her boyfriend, Rick Spero. She said she had given him a blue Cadillac El Dorado. The light blue van was hers. Other times she mentioned a mobile home located somewhere near Bakersfield, a ranch in Aqua Dolce, and a house in Lakeview Terrace.

"I think Barbara's got plenty of money," Cindi said.

Daylight sifted through a jagged edge of cloud. "Looks like Mama Dog's about ready to pop," Terry said. "Maybe it'll be today."

They had given Mama Dog her own run at these last stages of her pregnancy. That meant tripling up in some cages, but the Collie-Lab was due any day.

Terry kissed Cindi's forehead. "Come on, baby, let's go back to bed."

The living room was still warm as they slid under the covers. The dogs were at it again, but Terry closed his mind to the racket and drew his wife into his arms.

"Not now, Charlie," he said.

But the cat had already pounced on Cindi's feet and was settling in, purring softly.

•　　•　　•

The following morning dawn broke in soft hues over the Malibu canyons, casting a pastel wash over terracotta cliffs. By 6:30, a big man and four big dogs stampeded out of an oceanside motel, piled into a Bronco van, and headed up to the canyons.

"Wiggles, cut it out," Norman Flint said, laughing as the pup fought to sit on his lap. Good-natured Bear let his brother trample over him.

Their parents, Cody and Fred, patiently gazed out the window. "We'll be there soon, guys."

Nearly every morning they all swam in Big Tuhunja. Part Labrador, part Greyhound, the dogs loved to run, and they preferred the canyons, with their open spaces and fast-moving rivers, to the beach.

The quiet appealed to Norman, too. He had a lot to think about. There was not much work around for a builder. Another week and they would be moving to their house in Reseda; maybe then things would pick up. At least he had kept his little family together, despite constant harassment from Animal Regulation in LA.

"You are only allowed three dogs, Mr. Flint. You've got to get rid of one of them."

It made his blood boil to think of Lieutenant Penia issuing him warnings and then fines. He, of all people, who grated Parmesan cheese on his dogs' chow because they liked the taste. And he spent hundreds of dollars each month on toys and treats. Now he was being charged like a criminal with violating LA Municipal Code 53.50.

Who did Animal Reg expect him to give up? How do you make that choice?

The dogs piled out of the van and tumbled over each other heading toward the river. Of Fred and Cody's litter of six, two pups stayed. "One little doggie followed me around, underneath my feet all the time," and Norman had named him Bear. Bear had a big look, as did Norman himself. And there was another pup, a sociable little guy whom Norman's Mom named Wiggles.

"Hey guys, catch!" Norman tore off his shorts and jumped in with his dogs. Wiggles, the winner as usual, delivered the stick. Bear wanted only a hug. Relaxed and content, Norman stretched out near his dog family, naked and prone to the air and sun.

"You are allowed three dogs only, Mr. Flint."

How was he supposed to choose?

.   .   .

Twenty miles away, in suburban North Hollywood, Ellen Hickman awoke in a sweat. Five months had passed since Joel's suicide, but she still had nightmares. He had made a bed in the garage, and turned on the car engine. Then he went to sleep.

The police found a radio playing classical music beside him.

Joel had tried to kill himself seven times before. And he had threatened to kill her.

To Ellen Hickman, living for twelve years under the pale of her husband's depressions and hospitalizations, her animals were her only solace. She could not save Joel, but she saved and found loving homes for dozens of cats and dogs.

Now, Ellen's life was in chaos. She was feeling very fragile, vulnerable. Another cat was too much to handle; there were already five. Yet it was painful to have to give up Calico. Last night, the anticipated loss of the pretty stray was more than she could bear. She could not stop the tears.

"Don't worry, Calico," Ellen whispered. "We are going to find you a wonderful home with people who love you. I promise."

Still, she thought about canceling the ad in the *Recycler*. Then she realized it was Thursday morning. It was too late.

•   •   •

As Terry had predicted, the groom room of the kennels was crowded with new arrivals. An early riser, Barbara had already removed their collars and IDs and hung them on the peg board. The dogs were barking, and the cats huddled in their cages. Terry and Cindi talked soothingly to them as they brushed and flea-dipped them, then attached their red, brown, or yellow plastic collars. To each collar they affixed a bronze-color ID tag with a set of numbers. "That's to show they are boarded animals," Barbara explained. "In case Animal Reg comes by to inspect."

Terry and Cindi jotted down a basic description of each animal. Later, Barbara would transfer the information to her files. Lord knew how she found anything in that office. Cindi had glimpsed the room as Barbara entered or left it. It was a maelstrom of papers.

The kennel phone rang.

"Budget Boarding," Cindi answered. There was a pause. "Hi, no she's in her office, but bring them down. We ain't going nowhere."

"Bob's coming down with the cats," she called out to Terry.

They had told their friends, Bob and Anita Burns, about Barbara's charitable work. The Burnses jumped at the chance to place the dozen stray cats they were feeding, and Barbara seemed delighted to take them. She even said she would pay Terry and Cindi for the referral. Five dollars for any cat and ten for each dog. Just what she paid Ralf.

Through the half-closed office door, Cindi heard Barbara's voice, soft and cheery, talking on the phone. Nobody was supposed to answer that line, which was attached to an answering machine.

Cindi was placing a new arrival in the cat room when Rick Spero came in. He murmered his usual, "Good morning" and headed toward the office.

Barbara told the Phillipses that her boyfriend had been a real estate broker. Despite the fact that Rick seemed very pleasant, something about him gave Cindi the creeps.

"He's doing just fine," Cindi heard Barbara say into the phone. "Wouldn't *you* be happy on a ranch in Canyon country? It's paradise for him." Barbara laughed. "You are very welcome."

Cindi did not recall anything about a ranch, but that adoption could have occurred before they came to Budget.

Barbara poked her head into the cat room. She was in her usual drab jeans, shirt, and knee-high rubber boots.

"Cindi, if Stephens calls———."

"I know, take a message. And, Barbara, the Burnses are bringing their cats over."

Barbara said something to Rick about a shipment of cats going about the end of the month, and that she was on her way out to meet Ralf. Cindi wondered how Barbara juggled all the men in her life.

"Well, looks like your girlfriend's gonna find a home," Cindi said to Charlie, who was reaching into one of the cages to paw a long-haired Angora. Maybe the Burnses' cats could get in on that shipment.

●　　●　　●

About the time Barbara was tossing cages into her van, Chuck Ransdell was talking with his wife in their Studio City home. "Barbara said Ammo's doing great, Leslie. I asked her if we could take Travis up to visit around Thanksgiving. She's going to see about it."

Maybe now he could think about replacing the windows Ammo had crashed through. But as Chuck, a former actor now builder, got into his pickup and headed for the expressway, he half expected to hear Ammo racing behind the truck, barking. Ammo was wild and crazy and he needed a lot of room to run—the kind of place Barbara had found for him. But the dog had become part of their family. He had to admit he missed him.

"If it doesn't work out in his new home, we'll definitely take him back," Chuck had told Barbara. But all sounded well with Ammo.

The 405 Freeway was already bumper-to-bumper. Chuck turned on the radio and settled in for a long, tedious ride. A ten-acre ranch in Canyon county. That lucky dog.

•   •   •

The week of November 11th was hectic. Within three days, Ralf brought thirteen dogs into the groom room. Big dogs. Shepherd mixes, Lab mixes, Huskies. Eight dogs on November 13 alone. Barbara told Terry to tie some of them outside until room was found in the kennels.

Already there were fights over territory and food. Mama Dog had lost her privileged position and her five puppies were huddled against her in a corner of a crowded run. Terry tried to keep the squirming family clean, but the poor drainage system left rivers of urine in the kennels. The puppies were weak. The few that tottered around on spindly legs fell into puddles of urine and diarrhea. But Barbara would not allow them inside.

"The man from up north," as Barbara called him, had found "placement" for some of the animals. She said a shipment of cats would go out first. But many of the cats had eye infections, and Barbara seemed on edge. At about eleven o'clock on the night of November 15, Cindi heard Ralf in the cat room. The next morning there were seven more cats in stacked cages.

On the morning of November 17, Ellen Hickman opened the door to her well-kept home on Camelia Street. She saw a handsome young man smiling at her.

"Hi, I'm Steve Jacobs."

"Oh, yes, yes. Come in." She had not expected someone so good-looking. He reminded her a little of Joel, that same kind of charm, a dapper dresser, but "Steve"—Ralf Jacobsen—was blond.

Ralf smiled warmly as Ellen introduced the cats. "It's a little hectic here right now," she said apologetically. "I've been through a difficult time." Calico came up and rubbed against her leg. "This is Calico, isn't she pretty?"

"Yes, very, just like her Mom."

Ellen felt herself blush. "And this is Sammy. They aren't kittens. Are you sure you want grown-up cats?"

"No kittens!" Ralf laughed. "My fiancée would kill me."

Ellen felt a stab of disappointment. "Oh, so you are going to be married?"

Ralf told her his fiancée was home most of the time, that she worked at home. He was from Germany, and while a child his parents traveled a lot. "My mom and dad were often separated. I don't want a marriage like that," he said. "I want someone to be close to me."

Ellen was impressed with his sensitivity. Ralf offered to hold the cats while she clipped their nails; he seemed to have a natural affinity for them.

"I'm so glad someone like you is taking them. I've had this fear about someone coming and then selling them into medical research. Do you know about that?"

"Horrible," Ralf said. "I can't imagine doing that to a poor animal." He went out to his car and brought back a large cat carrier.

"Well, you are prepared! Nobody's come with their own carrier before," Ellen said, surprised.

As he put Calico and Sammy into the cage, Ralf said, "If you want, maybe I could take the other cats off your hands."

"Six cats! I wouldn't do that to you and your fiancée. Cats are expensive to take care of, and they need a lot of attention."

"Money is no problem and, like I said, my fiancée is home all the time."

"No, I don't think so. These are my babies."

Ellen asked him to write down his phone number and address.

"You can come visit us anytime. We have a lovely home," Ralf told her. He hugged her good-bye.

•   •   •

When Cindi first saw Ralf at the kennels in late November, she instantly disliked him. They had not heard his car pull up, but the racket the dogs were making could have muffled even an explosion. He was in Barbara's office, the door half open, and looking through a *Recycler* when Cindi walked out front.

The night before, Cindi had heard Barbara and Ralf arguing. "I told you not to bring in those kind of dogs!" Barbara shouted. "I said I'd only pay for certain types."

Barbara called Terry into the kennel. "Shave those Afghans," she ordered. "Shave them both smooth."

What does it matter if they're Afghans? Cindi wondered then. Afghans weren't too smart, but they made good pets.

Ralf acted like he was God's gift to womankind, Cindi later told her husband. "And don't you think it's weird, what Barbara said about Ralf leaving suicide notes for her to find?"

"He's a strange guy," Terry conceded.

"And what about him picking up strays?"

"Does she want him doing that?"

"She didn't say anything when he told her. I know she said she wanted him to go through the "Free to a Good Home" ads, people looking for good homes for their pets, which would make sense since that's what we're doing—finding them good homes. But she really didn't seem to care if he took strays too."

"Maybe she figures people aren't taking care of them, so they might as well go to better homes."

But it made her sick the way Ralf treated the dogs, kicking them, using that electric cattle prod on them. Ralf, she decided, was a mean son of a bitch.

The day of the cat shipment, Barbara was testy. The figures she was tabulating on a clipboard were not working out right. Too many of the cats were sick. They would never be approved.

"Terry," Barbara said, "while you're out in the kitchen ask Cindi if she knows where Charlie is."

THREE

Betrayal

"We had no reason to suspect Budget Boarding. They said
they were not selling to research."
—*Lt. Robert Penia, Department of Animal Regulation
City of Los Angeles*

December 1987–January 1988
County of Los Angeles

Animal Regulation for the City of Los Angeles was responsible for
maintaining the city's animal shelters, inspecting and licensing kennels, and enforcing anticruelty laws. Between July 1, 1985, and June
30, 1986, "Animal Reg" granted sixty-nine permits for dog kennels
which passed yearly inspection. Among the new permits was one
issued to Barbara Ann Ruggiero, who filed an application on January
3, 1986, to run Budget Boarding, a boarding and grooming kennel
then managed by Sheila and William Gordon.

When Officer Lisa Goodman from the Department's East Valley
Shelter inspected Budget Boarding, she observed that Barbara Ruggiero seemed "a very happy person, happy-go-lucky, but a bit flippant
about business procedures. She was more interested in flirting with
the truckers and the California Highway Patrol."

A year later, on February 4, 1987, Barbara applied for a permit as
sole owner of Budget Boarding Kennels, a business which she had
recently purchased for $8000 cash from the Gordons. Included in
Barbara's purchase of the business was Budget's name, advertise-
ments, inventory, and list of clients.

The Gordons had been leasing the site since 1984 from property
owner Bridgette Jacobsen for $1000 a month. The unauthorized lease
transfer to Barbara was to fuel a rabid hatred between the two women.
In her March 21, 1987, report, East Valley Inspection Officer Kathy
Schamber described conditions at Budget Boarding as "very good,"

and the kennel's new owner was granted a permit on March 25 to board a maximum of eighty dogs and twenty-five domestic cats.

On December 8, 1987, Barbara applied for a second permit to operate "Budget Boarding 2" on Norris Avenue, then the site of Comfy Kennels. The permit for "Budget Boarding 2" was expected to clear by year-end.

As Barbara's business flourished, East Valley Animal Shelter was dispatching its twelve inspection officers on a mission which occupied much of their time: combing neighborhoods for violations of LA Municipal Code Section 53:50. That regulation stated that any resident having more than three adult dogs or three adult cats must apply for a kennel permit.

Section 53.50 brought "Mike Rogers" to Lull Street in Reseda on December 3, 1987.

Officers from the West Valley Shelter, which monitored the Reseda area, had already begun issuing fines to Norman Flint in his new home. Norman conceded that something must be done. The choice was painful, but he decided that Wiggles, the more adaptable and gregarious, would have an easier time adjusting to a new home. Parting with Bear, his little buddy, was out of the question.

Through word of mouth, Norman began interviewing prospective owners. He decided on a nice young couple who lived in Sunland. Several hours later, he called. How was Wiggles doing? Not well, they said. He had already found a way of getting over the fence. Norman knew the couple lived near a busy road; Wiggles could get hit by a car and killed. So Norman brought him back home.

Norman decided he could not separate the brothers. Maybe together they would settle in and be happy in another home.

Now, in his living room, hugging and kissing Wiggles and Bear, was "Mike Rogers." It had taken nearly twenty interviews of people answering his "Two Lab Pups Free to a Good Home Only" ad to find Mike Rogers, the perfect "parent" for Wiggles and Bear.

He told "Mike" that Wiggles could easily scale a six-foot fence. A block wall would be ideal.

"Then you don't have to worry," Ralf said. "We have an eight-foot-high block wall. No way they can get under or over that!"

"Will you have time to play with them?" Norman asked.

"My wife is home, which is why I want company for her. I think the dogs will be getting more attention than I will."

They laughed as Wiggles and Bear tried jumping into Mike's lap. "These guys are great with kids," Norman said.

Ralf said, "My wife is expecting."

It sounded perfect.

An hour later, Wiggles and Bear bounded out the door, chomping at each other's legs and ears. Wiggles jumped onto the back seat. But Bear held back.

"Go on, buddy," Norman said. "I'll be seeing you real soon." As he eased Bear into the back seat, Norman felt tears gathering behind his eyes.

"Bear's really my buddy," he explained.

"I know, there's always a dog like that. I had one when I was a kid I'll never forget. And hey, you can come pick them up to go swimming any time."

Norman clasped Ralf's hand. "Listen, man, thank you."

"Don't worry Norm," Ralf said. "They'll be fine."

Norman controlled his impulse to call. He waited two weeks and then, on December 15, he dialed the number "Mike" left on a piece of paper along with his address: 5565 Corbin, Woodland Hills.

"The woman who answered said I had the wrong number," Norman told his mother when she asked about the dogs. "She said no Mike Rogers lives there. Maybe he didn't want me to come take them. He seemed to get attached to them right away."

Still, maybe they should pass by the house, just to ease his mind.

Heading north on Corbin, Norman and his mother kept an eye out for number 5565. He stopped the car at the spot where Mike Rogers said he lived. Had there been a 5565 Corbin, it would have been on the ramp of the 405 Freeway.

That same morning, Paul Iverson decided to check up on PJ. Not that there was anything to worry about. "Steve Jacobs" had been down on his hands and knees playing with PJ. He said he wanted a companion for his wife, who spent a lot of time alone on their ranch; he was gone two weekends a month as a pilot in the Army reserves.

Paul explained that, both working and going to school, he did not have enough time for PJ. "Steve" said he could come by to visit PJ whenever he wanted.

Paul decided to take Steve Jacobs up on his offer. He found the paper with Steve's address and home phone and dialed the number. A young woman answered, "Blue Cross of Southern California."

Something told Paul he had not misdialed. He slammed the phone down, jumped into his car, and headed from Simi Valley toward Woodland Hills and a house at 6134 Winnetka that did not exist.

As Paul Iverson made his way to Winnetka, the blinds in Ellen Hickman's North Hollywood bedroom were still drawn. She had not slept for days. Since November 20, she had been unable to do anything but cry. She had called Steve Jacobs, but it was the wrong number. She had driven to his house; it was a vacant lot. Two weeks later, a "Steve Jacobs" again telephoned looking for year-old cats. "Are you the same Steve Jacobs that came here?" Ellen managed to ask. "Steve" hung up. Then, a "Barbara" left a message on Ellen's machine. She was interested in the German Shepherd Ellen had advertised.

Who were these people? What were they doing?

On this gray morning, crumpled loose-leaf pages lay in a pile at the foot of Ellen Hickman's bed. On each she had written the question that tormented her day and night:

Steve Jacobs, where are you? Where are my cats?

• • •

"The man from up north" was expecting three more deliveries from Budget. Dogs and cats had already gone out in the first shipment of December. The morning of December 15, Cindi and Terry were in the throes of cleaning and packing for the second shipment: cats only.

Terry and Cindi had been up half the night flea dipping and checking eyes and ears for any sign of the infection that swept through the cat room. The groom room was still packed with the morning's arrivals. Dogs were more than five deep in the chicken coops. It was virtually impossible to control the feces and urine that accumulated within hours. Cat cages were stacked four high along two walls, with several cats in each cage.

Rick Spero had gotten hold of the lock-box keys to their new kennels on Norris, and though it was not yet officially theirs, he and Barbara gained access. Terry and Cindi were told to clean and disinfect that kennel, but Barbara warned them not to be seen on Norris walking to or from the kennel.

Several days after the second shipment, a frost settled in. One morning, Terry found two of Mama Dog's puppies frozen to death in their own urine.

"Put them in ziplock bags in the freezer," Barbara told him. "You can defrost them later for Monty."

* * *

That same evening, two hundred yards from Comfy Kennels ("Budget Boarding 2") Candy Sheker lay sleepless in her bed. The strange part about it, she thought, was that Pooches would never have gone off without Kasey or Junior. Those two Shepherds were very protective and aggressive. They treated Pooches like a son; they would never have left him on his own.

The panic which had gripped Candy two months before raced through her veins like cold silver. That day, when she returned from work and found the yard gate open, she had driven to all the city shelters on the chance someone might have found her dogs on the street. She went through the "Found" ads in the *Recycler*. She put her own ad in the "Lost" section. She handed out hundreds of fliers with a photo of Pooches, Kasey, and Junior at every kennel in her neighborhood. Had anyone seen a young black Lab and two Shepherds? They were well-fed, well-groomed. They were not roamers. They had never gotten out of the fenced yard.

Three weeks into her search Candy Sheker received a phone call. Two Shepherds had been sighted roaming up in the canyons. Someone had seen a car drop them off. Now Kasey and Junior were safe at home, but there was no sign of Pooches.

Candy Sheker, a big, blond, straight-talking woman, heaved herself from bed and walked to the window. Kasey and Junior were close at her heels, their ears were cocked to the familiar sound.

She'd know that bark anywhere: a peculiar sort of high-pitched howl. Sometimes she'd hear it in the morning, sometimes late at night. The first few times she ran out of the house and down the street like a mad woman shouting "Pooches, Pooches!" The barking seemed to get louder, and then it would abruptly stop.

She had never behaved like this in her life. Everyone knew her as down-to-earth, practical. Now her husband, Rick, thought she was going crazy. Maybe she was. But she did have someone who believed her, who was willing to help. Comfy Kennels' new owner seemed genuinely concerned when she heard about the missing young Lab. She promised she would keep an eye out for Paw-Pooche.

"If I see any dogs loose on the street," Barbara Ruggiero told Candy Sheker, "I always bring them in."

•   •   •

As Christmas approached, Animal Regulation Officer Lieutenant Bob Penia received a call at the East Valley Shelter. Barbara Ruggiero told him she had overbooked for the holidays. Could she move some of the animals to the kennel she was purchasing on Norris? Penia, a soft-spoken Hispanic, had been with the Department of Animal Regulation for the City of Los Angeles for fifteen years. He had been attracted by the pay scale and taken a job on a work program. He ended up a career inspection officer. The father of three and owner of several dogs, Penia loved working with people and animals.

In late December there were, on any given day, about 150 animals at the East Valley Shelter. Some of these dogs and cats had been found on streets, abandoned, or given up for adoption. The shelter was not particularly aggressive in its adoption services. It adopted out or returned to owners less than 20 percent of those animals, about the national average for animal shelter turnovers.

And, like most shelters, East Valley was not actively promoting its low-cost spay/neuter clinics designed to curb the pet population. Each year, that shelter alone euthanized about 10,000 unwanted or unclaimed pets. The yearly cost of destroying these animals was over $100,000. Critics of the Department claimed that if those funds were directed to an aggressive spay/neuter program, the population problem would diminish, and with it the numbers of animals euthanized.

Each day East Valley averaged ten calls from residents reporting missing dogs. These were predominantly Shepherds, Huskies, and their cross breeds. Over 3500 dogs were reported to East Valley Shelter as missing each year. Officers attributed those losses to roaming coyotes or road kills, or to the familiar axiom: Finders Keepers.

Pet theft was always a possibility, but nobody at the Department seemed to seriously consider it. In its nearly thirty-five-year history, the LA Department of Animal Regulation had generally disregarded calls alleging pet theft.

During 1987, Animal Reg inspectors had reported on Budget's poor sanitation, sick animals, plumbing problems, and the usual general clutter, and the presence of a snake that was unlicensed. Still, the kennel was recommended for licensing, and Lieutenant Penia decided there was no reason to deny Barbara's request to accommodate the Christmas rush at Comfy.

During the holidays, Linda Gordon, a senior Animal Control Officer, stopped by Von Konstanze Kennels. Linda was not there on official business. She was interested in buying one of Von Konstanze's prize German Shepherds. Gordon was familiar with the history of animosity between Bridgette and Barbara and usually ignored their respective complaints. But today there was something about Bridgette's vitriolic rage that caught Linda's attention. The kennels, Bridgette said, were filthy. There seemed to be hundreds of dogs and God knew how many cats. The stench was unbearable. Barbara was letting the place deteriorate beyond repair; the plumbing was already shot.

Officer Gordon had met Barbara in late 1984, shortly after Delores Ruggiero had been killed. Barbara's mother had been bludgeoned and strangled to death in her own bed by an unknown assailant. With the money she inherited, Barbara had purchased Budget Boarding.

Gordon decided to have an unofficial look at the kennel run by Delores Ruggiero's daughter. She peered over the concrete wall that separated Von Konstanze from Budget.

According to California Animal Reg penal codes, any animal in a kennel like Budget Boarding which goes without care in excess of twelve hours was considered abandoned. By the appearance of the kennels, the piles of feces and apparent ill health of the dogs, Budget Boarding looked abandoned. Officer Gordon filed a complaint at the East Valley Shelter claiming that Barbara Ruggiero had abandoned her facility.

On January 1, Officer Nancy Caump was dispatched to Budget Boarding. She reported that the sanitation was indeed poor and that four cats had discharge from their eyes. Officer Caump observed medication which the owner said was being given to the cats. She issued a Notice to Comply to Barbara Ruggiero, instructing her to contact the supervisor of the East Valley Shelter concerning her efforts to clean up the feces and overall unsanitary condition of her kennel.

At 3:50 p.m., on January 4, Officer Eric Gardner revisited Budget

Boarding. The sanitation had improved but there were still two sick cats. A followup call was made several days later by Officer Gardner. No one was on the premises.

• • •

At 6:30 on the evening of January 6, as yet another shipment of dogs left Budget Boarding, the case that would be known as People's Number A713152 took a fateful turn. That evening, Sheri Lamotta* opened the door of her sprawling contemporary home in suburban Tarzana, with its English perennial gardens and pool. Her first thought was that this "Mike Johnson" was, fortunately, blond.

"What are you going to say in the *Recycler* ad, honey?" her husband, Rob, had asked her jokingly a week earlier. "Blond people only, for English Springer Spaniel?"

The situation had indeed gotten out of hand.

Their dog, Maxwell, hated anyone with dark hair. That put her husband, Rob, and her housekeeper, Carmen, at high risk. Several days before, Maxwell had pinned the maid against the living room window and the glass had shattered against her back. Lately, Max would not even let Rob into the kitchen.

Only Sheri, a strawberry blond, and her fair-haired children were safe. Now Sheri watched this very charming young man cuddle Max. He had an easy manner that made her feel she was doing the right thing.

"Spaniels are my favorite dog," Ralf said. "I had one like this when I was growing up."

"Where was that?" Sheri asked.

"I'm just a California country boy," he said. "I've got a ten-acre ranch in Aqua Dolce. Max will love it there."

"What kind of work do you do?"

He told her he was a pilot, and that he going to law school at night.

Sheri Lamotta assessed people largely on their appearance. "Mike Johnson" was clean-cut, well taken care of. He would take good care of Max. Still, it broke her heart to give Max away.

"Is Max very attached to your other dog?" Ralf asked.

"I guess he is."

*Name changed at request of Subject.

"Why don't I make it easy on him and take Adrian, too? That way——."

"No." Sheri shook her head. She had already begun to cry as she gathered Max's leash, his terracotta bowl, the big pillow that was his bed. She asked "Mike" to leave his address and phone number, just so she could make sure Max was okay.

Rob offered to take Max and his paraphernalia to Ralf's car.

"You know, it's strange," Rob said when Ralf had driven off with Max. "That guy had another dog in the car, and there was a cage. And he was driving a Mazda RX7, which is a weird car if you're living on a ranch. Sheri, where are you going?"

Sheri would later recall racing to her car and crying over and over again, "Oh, my God, please God, no." It was a gut feeling, something she could not explain. All she knew was that something was very wrong. Waves of icy panic numbed her fingers on the steering wheel as she sped around the neighborhood looking for Ralf's car.

At eight o'clock, Sheri returned home, exhausted, shaken. She went into the kitchen and began calling the other "Free to a Good Home" ads in that week's *Recycler*. This "Mike Johnson" had taken at least two dogs. He might have taken more. Someone may know where to find him.

Half a dozen pet owners told her that a clean-cut man in his twenties named Mike Johnson or Steve Johnson or Steve Jacobs had adopted their dog or their cat, often taking with him more than the one animal they had advertised. They had all been impressed with the young man's charm and sincerity.

They had given him leashes, toys, even Christmas stockings for their pets, said tearful good-byes to Shepherds, Labs, Huskies, Tabbies, and Calicos and, overall, felt guilty about the decision they had made. Now they had no clue as to who this guy was. But they all wanted their dogs and cats back home.

At eleven o'clock that night, Bill Dyer, a playwright and concert organizer, received a call from an hysterical Sheri Lamotta. The police had not responded to her call, nor to the subsequent calls of other victims. She had been told by her mother that Dyer worked with communities concerned with protecting their pets.

Could they help her find Max?

Bill asked Sheri if she had heard about "dealers" or "bunchers." She had not.

Bill explained that these people buy and sell animals—dogs and cats—for laboratory research use. Class A dealers bred animals for that purpose. Class B dealers obtained animals from "random sources," meaning from just about anywhere—including the "For Free" ads, even streets and yards. There was no enforcement of any regulations by the government agency that licensed these people. He told her that both classes of dealers were licensed by the U.S. Department of Agriculture.

All Sheri knew about USDA was that it stamped the cellophane packages of meat in supermarkets.

Bill explained that USDA has significant tasks other than inspecting and grading meat. It's budget reflects its enormous scope. The fourth largest federal agency, USDA has an annual outlay of about $62 billion. Viewed on an international scale, only seventeen nations in the world have budgets that exceed that of the USDA. In budget and spending, the agency would rank fourth among U.S. corporations, smaller than General Motors, Exxon, or Ford but larger than IBM, Mobil, and General Electric. Based on its credit and lending activities alone, USDA would be considered one of the largest banks in the nation.

Over half of USDA's budget goes toward food assistance in the form of Food Stamps, child nutrition programs, and other public assistance programs. About 40 percent of its workforce is employed by the U.S. Forest Service to provide forestry services and outdoor recreation facilities. The agency has a housing loan portfolio of nearly $30 million, and it is a major factor internationally, with personnel in most embassies and over $8 billion annually in export programs.

Unknown to most people, USDA also oversees research programs nationally and world-wide. And it is responsible for enforcing the Animal Welfare Act—the 1966 legislation created in an attempt to stop the rampage of pet theft for sale into medical research. As such, USDA licenses "dog dealers" and their kennels, as well as institutions using animals for research. The law that USDA is mandated by Congress to enforce includes standards for the procurement, care, and handling of dogs and cats for research purposes.

But twenty years after the Animal Welfare Act had passed, pet theft was still occurring in epidemic proportions. USDA was simply not enforcing the law. Neither was it cracking down on the hundreds of thousands of unlicensed "bunchers" who supplied dealers with stolen pets.

Bill told Sheri that chances were this "Mike" character was a buncher.

Max sold for medical experimentation? Sheri felt faint, and then hysteria took over. She heard Bill say that pet theft was related to the selling of dogs and cats from the pounds to biomedical research laboratories.

"What do you mean?" Sheri screamed. "If they've got enough pound animals, why should they take ours, from our home?"

Sheri Lamotta was about to get her first lesson in the pet theft racket. The substance of that conversation would change her life and the lives of hundreds more victims in the Los Angeles area.

•   •   •

At the time of Sheri's call, the animal-protection community and local residents were fighting a decades-old battle with the San Bernadino City Council. The Council was requiring its city pound to sell dogs and cats to medical research. Each year the San Bernadino Pound sold about 1000 dogs and cats to the Jerry L. Pettis Memorial Veterans Hospital, Loma Linda University Hospital, the University of California at Irvine and Riverside, Harbor/UCLA Medical School, University of Southern California Medical Center, Cedars Sinai, and the Veterans Hospital in San Diego.

"The animals (in research facilities) seem to be treated better than most people treat their own," San Bernadino Mayor Bob Holcomb told the local press. "If I were the dog, I'd much rather go to research."

That year, several states required that laboratories be given access to unclaimed pound animals. Iowa, Minnesota, Ohio, Oklahoma, South Dakota, and Utah actually mandated "pound seizure"—that is, the required sale of animals from the pound to research. Meanwhile, Connecticut, Delaware, Hawaii, Maine, Massachusetts, New Hampshire, New Jersey, New York, Pennsylvania, Rhode Island, South Carolina, West Virginia, and Vermont prohibited this practice, although many of those states still allowed the sale of cats to research. Residents in the states that banned pound seizure had voted that an economic relationship between researchers and animal shelters was a violation of public trust in a tax-supported community service.

The remaining states allowed researchers to obtain pound animals under certain conditions, or they left that decision to local jurisdictions, as in San Bernadino, California.

Opening pounds to researchers was a highly emotional topic. The animal rights movement had made the public aware of the ethical issues posed by the use of animals in science and industry. But there were also coalitions of prominent health care professionals who were questioning the scientific value of using animals, and, even more specifically, animals from pounds. Many of these dissenters were at loggerheads with their peers whose salaries depended on grant money for animal-research projects. Taxpayer-supported grants for animal-based research totaled about $5 billion in 1987, $9 billion in 1991.

Now legitimate issues were being raised by people like Sheri Lamotta about just whose animals were ending up in pounds and laboratories.

Animals in pounds are former pets, either lost, abandoned, or given up by their owners in the hopes that the pound or shelter will find them good homes. These dogs and cats, which were "too aggressive" or "too gentle," "too big" or "too small," simply had the misfortune of being subject to their owners' whims. Some will be among the 60 million animals used yearly in research from invasive procedures like cancer therapies to cosmetics and toxicity tests for household products, and classroom dissection.

Expectant of reward and human affection, loyal and trusting, "companion animals," also known as "random-source animals," are considered ideal in laboratories since they are accustomed to being handled by people. An eyewitness account of a stress experiment using companion dogs at the University of Minnesota showed how accommodating they are. Thirty-six dogs were tied up and given electric shock from which they could not escape. Then they were shocked again. But given the opportunity of escape, the dogs did not even try. They just whimpered and looked "pathetically helpless."

Like most research facilities, the University of Minnesota obtains dogs from its state's pounds and from dealers who buy from bunchers, one of whom was convicted of pet theft. It also obtains dogs and cats from Midwest dealers buying at auction, and it accepts "gifts" of dogs and cats brought right to its laboratory door.

"Bunchers and dealers consider any animal on the street or in a yard fair game," Bill told Sheri. "They don't chose the sick or injured ones. They want healthy, socialized animals, often specific breeds, and the best way to guarantee that is through the "Free to a Good Home" ads."

"But what I don't understand is, with so many animals killed in pounds, why do they need to steal more?" Sheri asked. All she could envision was Max lying prone on a laboratory table. Why did it have to be her dog? Why not some stray?

Sheri was asking a common question. Researchers have long argued that allowing pounds to sell to laboratories would meet their demands and reduce alleged instances of pet theft. However, experience proved otherwise. In 1966, most pounds in this country were open to researchers. Still, that year, *Life* Magazine disclosed that two million pets were stolen and sold into medical research. The article prompted the passage of the Animal Welfare Act, which the USDA was required to enforce, to specifically stop pet theft.

Hearings in Congress in 1966 over the Animal Welfare Act conveyed to the public the enormity of the pet-theft problem. The bill's sponsor, New York Congressman Joseph Resnick, told the committee: "The animal procurement practices of the nation's research laboratories have become a national disgrace. Family pets, dogs and cats, are stolen off the street, sold to disreputable dealers, and eventually wind up in the hands of suppliers to hospitals and laboratories."

"What people have to realize," Bill now explained to Sheri, "is that research is a business, an industry, and dogs and cats are just the tools that qualify researchers to apply for grants. Grants need to be looked at for what they are: salaries paid by us, the public, for work we are not allowed to examine critically."

As in any business, supply and demand dictate, and the demand from laboratories can be very specific. In 1984, one institution in Cincinnati needed quantities of poodles; soon pet registry hotlines buzzed with missing poodles in the area of that lab. In 1986, a Pennsylvania lab put out an order for one-hundred twenty-six German Shorthaired Pointers. When that breed began disappearing from nearby counties, the source of research dogs was clear.

Bill told a horrified Sheri that in states where pounds legally sold to research and dealers, neighborhood pets were not safe. Documentation accumulated by Mary Warner, founder of Action 81, the national pet-theft tracker, from 1979 Virginia police reports, lost pet registries, Animal Control Officers, dog and cat clubs, and victims showed that when out-of-state dealers had access to Virginia pounds, over 8000 dogs were reported missing in "areas in proximity to those pounds" within one year. In 1980, when out-of-state dealers were barred from Virginia pounds, pet-theft numbers dropped by 40 percent; theft dropped by 50 percent the following year.

Dealers thrive in areas where pounds legally sell to research. Some of the richest dealers operate in Minnesota, Iowa, and Oklahoma, where they can buy at pounds. And when supplies of premium dogs (Huskies, Labs, Shepherds, and cross breeds) are low, bunchers hit the streets to supplement. Minnesota dealer Don Hippert, who also ran the Dodge County pound, bought from such a buncher. The result: on a tip from a Mower County sheriff, sixteen pet owners found their stolen dogs at Hippert's clients—the Mayo Clinic and the University of Minnesota. Hippert's buncher had stolen them from their owners' property.

The information overwhelmed Sheri. What Bill was saying was that researchers were knowingly buying stolen pets. How could they get away with that?

Bill said, "Researchers blame theft on USDA, which is supposed to check dealer kennels and records but doesn't. But researchers know USDA is not doing its job. If it did, the supply of cheap dogs and cats would diminish, and prices would rise."

Bill explained the enormous financial stakes. The medical industry constitutes 15 percent of the country's Gross National Product. In 1988, when the National Institutes of Health, the principal funding agency for medical research, granted over $5 billion taxpayer dollars to research projects, 44 percent or $2.3 billion of that money funded experiments using animals; only 23 percent, $1.29 billion, went to clinical studies of human patients.

Sheri said she knew rats and mice were used in research, but what were researchers doing with dogs and cats?

Bill ran down the list: coronary experiments in which heart attacks were induced in dogs; cancer and lung disease experiments sponsored by the tobacco industry in which dogs were forced to inhale cigarette smoke; psychological stress tests in which kittens were separated from their mothers to study the stress of maternal deprivation; ocular experiments in which kittens' lids were sewn closed to judge the effect on their motor reflexes; radiation tests sponsored by the Defense Department in which dogs were burned alive; head trauma experiments which involved shooting thousands of cats and dogs at close range. These experiments and classroom exercises were, according to a growing number of physicians, often redundant, lacked scientific merit, and simply substantiated what clinicians working with human patients have known all along. In other words, they were unnecessary.

"But isn't any of the research important? I mean, it can't all be so cruel and ridiculous," Sheri protested.

The medical profession's divided opinion of the scientific validity of using random-source companion animals became apparent in 1984 when attempts were made to stop California pounds from selling to research. Sheri vaguely recalled something in the press about the issue, but it seemed irrelevant to her life. She had not connected her own dogs with those pound animals; certainly not with animals in laboratories.

Bill filled her in on what was going on behind the scenes. During that campaign, over seven hundred physicians signed a strong statement critical of the "poor science" of using random-source dogs. That statement read in part:

> Pound seizure is an ill-conceived practice damaging to the good name of science and to its quality. The use of animals from shelters for experimentation is not only unnecessary and unethical but it is detrimental to sound research. (These animals) are of undetermined genetic, environmental, and medical background. They react unpredictably and inconsistently, making questionable the reliability of most research in which they are used.

For those reasons even the World Health Organization publicly advised against the use of pets in medical research, as did the Council of Europe. Because of their questionable value in research, the use of dogs and cats from pounds is banned in Sweden, Holland, and Denmark.

Using random-source dogs was not only scientifically questionable but very expensive, more so than using "purpose bred" dogs which initially cost more. Numerous studies by research labs including the National Institutes of Health have found that *double* the number of random-source dogs is often needed for experiments. The reason: a high death rate. Former pets cannot endure the stress of laboratory conditions.

But cost did not appear to worry researchers in California, where the measure to stop pound sales to research failed. As Dr. Julia Bailey, former Administrator of the University of Southern California Medical School Research and Training Center, wrote to then California Senate president David Robertti, "Cheap pound animals may permit higher salaries for researchers."

In 1981, the residents of the City of Los Angeles denied researchers access to pounds. The City decided "pound seizure" violated the purpose of shelters and pounds, which is to return lost animals, adopt out others and, if all else failed, humanely euthanize these once devoted companions. Putting a price tag on pound animals would endanger the safety and well-being of all pets by making them commodities.

Yet the County of Los Angeles allowed the "qualified" sale of pound animals to research. In fact, the County was bending over backward to help its research partners. Brian Berger, in 1988 the director of LA County's Animal Care and Control, was quite willing to favor researchers. ". . . There must be a reasonable way to make animals available to you by the more general interpretation of the word 'pet,' " he wrote to Dr. Jesse O. Washington, director of the Division of Laboratory Animals Medicine, UCLA.

Berger's plan was to set aside a designated day of the week for each institution to come discreetly to the pounds and buy.

A personable twenty-one-year veteran of the Department, Berger was eager to please. He suggested an "exchange program so that on any day if there is more than you need, they could be distributed to someone else." Researchers could thus stockpile preferred animals— those that are gentle, healthy, easy to handle. In effect, the most adoptable animals which, given time, might have found homes were reserved for researchers.

Pounds in the County of LA were, meanwhile, riddled with suspicious activity. In 1983, an audit of Animal Control showed that in one month alone two hundred dogs and cats were "missing" from at least two county shelters. Twenty percent of those dogs and cats were supposedly put to sleep, but euthanasia records did not substantiate this. Another 13 percent were supposedly residing at the shelters, but they could not be located.

The problems in the County of LA were magnified a hundredfold nationwide. With billions of dollars at stake and the need for discretion, alliances have formed between city councils, law enforcers, pounds, and researchers across the country to ensure the flow of cheap, easy-to-handle pets into laboratories. These relationships were transforming pounds and shelters into laundering waystations for stolen animals.

The pounds in Norfolk, Virginia, for instance, have set aside areas inaccessible to residents looking for their "lost" pets. In these restricted areas, premium "lost" or "unclaimed" dogs were reserved for researchers. In one instance, in April 1982 an "unclaimed" Shepherd was sold by the pound to the Eastern Virginia Medical Authority Laboratory in Norfolk. On a tip from a pound worker, the frantic owner, accompanied by a local television crew, went to the lab. One member of the crew was bodily ejected, but the pet owner recovered her dog. The Shepherd had been shaved in preparation for experimentation. Days after bringing him home, she noticed metal passed in his stool. Her dog had already been used for ballistics tests—the metal objects were bullets.

Unscrupulous pound managers in certain counties in Missouri ensure that pet owners never recover their animals by delivering them directly to dealers or research institutions. One pound manager in that state actually responded to "Found" ads in newspapers to ensure a steady supply to his local research clients. He simply claimed the lost dogs as his own.

Some pounds actively discourage adoptions by offering price breaks to laboratories. In Minnesota, where researchers can legally buy dogs and cats at the pounds, the University paid $14 per dog while residents interested in adopting were charged over $40. In many counties in Minnesota, Wisconsin, Pennsylvania, and Virginia, dog dealer kennels *are* the pounds. Some government employees triple their salaries by bunching or dealing unclaimed or stolen pets. The obvious occurs: few "lost" Shepherds, Labs, Huskies, or Retrievers are ever returned to their owners.

Eyewitness and media accounts in Missouri, Iowa, and Arkansas describe Animal Control officers stealing dogs and cats off the street for multimillion dollar dealers who warehouse thousands of these pets in anything from front-load dryers and rabbit hutches to football-field-size kennels.

"We got dogs coming in with their damn ID collar on. Nobody says anything," explained one lab technician at an Arkansas research facility that has a contract with a major licensed dealer. The technician asked not to be identified for fear of reprisals. "It's common knowledge people can get $25 dollars for any dog they bring in, doesn't matter where it comes from. We've gotten in Irish Setters, Cocker Spaniels, all pedigrees. They just toss the collars into a heap."

●   ●   ●

The reality of pets in laboratories had hit the Lamotta household. By ten o'clock the following morning, January 7, Sheri Lamotta and a score of enraged LA residents determined to reclaim their animals had mobilized. The victims' sentiments were reflected in Sheri's comment: "We never thought anything like this could happen to our families. We had to do something."

FOUR

To Catch a Thief

Adult dogs—41–74 pounds, $350.00.
75–99 pounds, $400.00.
100+ pounds, $500.00.
All cats—$100 each.
—*Price list*
Barbara Ruggiero, "Biosphere"

January 1988
City of Los Angeles

A legal secretary, a playwright, a bookkeeper, waitresses, housewives, and other victims gathered back issues of the *Recycler* and the *Daily News* and they hit the phones with questions that were bizarre and terrifying. "Hello, my name is Sheri Lamotta. I'm calling because somebody has been obtaining people's pets illegally and, we believe, may be selling them to research facilities. Did a Mike or Steve Johnson or Steve Jacobs come to your house?"

The calls hit a nerve. Only one person, a man named Jack Plante, hung up angry at Sheri. More victims joined in the telephone blitz. Connections were forged among strangers in a phenomenon peculiar to victims who share a traumatic experience. The most profound feelings of the human spirit—grief, helplessness, rage, guilt—were almost palpable through the telephone lines.

These raw feelings were exacerbated by the reaction of law enforcers. Nearly all the victims had contacted their local police. The matter of "pet napping," as police termed it, was almost uniformly laughable. Several victims phoned the District Attorney's office in San Fernando. They were unable to clear the receptionist's desk. As for calling Animal Regulation, most residents felt they had only been harassed by the city agency. Many victims blamed LA Municipal Code 53.50, which restricted the number of pets, for forcing them to relinquish their animals to this scheme. The city's "barking" ordinance

41

had become a convenient tool for feuding neighbors, and there were allegations that the shelters had mistreated animals.

In one case Animal Reg ignored multiple warnings about an elderly Valley resident who for fifteen years had been "adopting" dogs from the "Free to a Good Home" ads. For a placement fee of $50 to $100 she promised their owners she would find their pets good homes. She earned thousands of dollars a month simply releasing the dogs and cats into city parks, or delivering them to middlemen who sold to research. In 1986, at the prompting of actress/activist Gretchen Wyler, a sting operation conducted by LA District Attorney Earl Sittel busted the woman in a late-night raid filmed by local Channel 2 News.

The victims of this latest scam decided to take justice into their own hands.

Unknown to most of these pet owners, their local hospitals and universities were centers for animal research that was funded with hundreds of millions in taxpayer dollars. Cedars Sinai Medical Center, also known as the "hospital to the (Hollywood) stars"; the University of California at Los Angeles; Loma Linda Medical Center (site of the failed Baby Fae/baboon heart transplant); City of Hope, and the Veterans Administration hospitals in Sepulveda and Riverside were some of the major users of dogs and cats in research. In 1988, Cedars Sinai received $5.6 million in grants for research from the National Institutes of Health. UCLA received $141 million for research from NIH; Loma Linda, $2 million; City of Hope, $3.9 million. The Veterans Hospitals in Sepulveda and Wadsworth received smaller amounts from NIH since the bulk of their research monies was distributed by the Department of Veterans Affairs. The VA allotted $192.9 million to its facilities for research that year. These same institutions also received millions of dollars more from other sources. For instance, Loma Linda's NIH grant represented only 15 percent of the twenty-three grants researchers there received. The institution also benefited from grants from the National Science Foundation, the VA Administration, the state of California, and private foundations. Likewise, Cedars compiled an additional $100 million endowment.

These universities and hospitals conducting animal research in the LA area purchased dogs and cats from county pounds and from USDA licensed Class B dealers. Many of these dealers became wealthy on a very low investment, as little as $40 for a USDA license. Each year they would pay USDA increasing fees based on their total sales. For sales of over $100,000, a usual base income for an average dealer, the fee to USDA would be $760 a year.

The requirements to become a dog dealer are minimal. The applicant must have a kennel that will pass inspection—no great feat since USDA was licensing dealers who kept dogs chained to rusted fenders in backyard junkyards or housed in clothes dryers. The kennel must have a veterinarian who can administer vaccines and a program of disease control, although in reality such a program is rarely undertaken.

With the substandard conditions persisting at dealers, even research universities complained to USDA about the health of the animals delivered. There is more than a 50 percent mortality rate for dogs and cats at many USDA licensed dealers. Half of the animals which dealers stockpile simply die before bringing in a cash return.

Once the applicant kennel passes inspection—and few are rejected—the dealer is given a booklet about the Animal Welfare Act's requiring minimum standards of care, sanitation and transport, and basic record keeping. Novices quickly learn that there are rarely penalties for failing to maintain any of these regulations other than a Letter of Warning, which is simply a slap on the wrist.

Los Angeles research institutions were supplied by several USDA-licensed West Coast kingpins: James Hickey and David Stephens in Oregon, Bobby Whitehead and Greg Ludlow in Arizona, and long-time supplier Henry "Bud" Knudsen in California—until Knudsen was busted by a county sheriff and charged with 124 counts of cruelty. What the San Joaquin sheriff saw that early November 1984 morning at Knudsen's kennels was something, he said, he will never forget: nearly one hundred starving dogs and cats foraging amid carcasses, wallowing in eight hundred pounds of their own fecal material. Some of these animals were later identified as "missing" pets or those that had been "adopted" from "Free to a Good Home" ads from as far away as Oregon and Washington. Knudsen's kennels had been one of the first licensed by USDA in 1967.

Conditions were also deplorable at S&S Kennels run by James Hickey. For years, USDA inspectors had found residents' "missing" dogs there and cited substandard conditions, but had taken no action. Citizen outrage prompted the local law to act in 1987, and Cedars' principal supplier was found guilty of "falsely reporting the dollar amount of his sales, failing to allow inspectors access to his business records, and showing false and misleading information about the description, number, and origin of the dogs and cats he acquired and sold to research facilities, thereby concealing the source of stolen and fraudulently obtained pets found on his premises." In short, pet theft.

As for Arizona-based dealers Ludlow and Whitehead, evidence was mounting for a class action suit by Greyhound owners whose dogs had been illegally sold to the Letterman Army Institute in San Francisco. These animals were slated for bone grafting experiments conducted by the Army and 3M Corporation.

But soliciting animals door-to-door was not the "kingpin" style. For the down and dirty, they had their bunchers, people like "Steve/ Mike."

•   •   •

At 5:30 p.m. on January 8, the phone rang at the Plante residence in Sunland. Jeannie Plante, a waitress at a fast food restaurant, was in no mood to talk. An hour and a half earlier she had fastened the short blue leash onto Butch's collar and said a tearful good-bye to the Golden Lab with the butterscotch eyes.

They had adopted Butch only a year before, but he had begun fighting with their Shepherd. Mickey was the one doing the provoking, but he was too old to place. The Plantes were forced to find Butch another home. Jeannie thought it was strange how Mickey went crazy when he saw Butch leave; he seemed to want to tear apart Butch's new owner. Now there was that crazy woman on the line, the one Jack said had called days before.

"Too bad if she can't get her dog adopted," Jack had told his wife. "It's like advertising anything, someone always wants to get the jump on you. You make sure you get this guy to take Butch."

"Look, Miss Lamotta," Jeannie said sharply to the caller, "he took our dog, okay? I'm sorry if yours didn't get adopted, but it wasn't right your trying to————"

"Excuse me. You gave your dog to that man Mike already?"

"Yes."

"Was he driving a blue Mazda RX7?"

"Yes. Why?"

"Oh, my God! This man is going around and collecting all these animals—ten, twelve, fifteen a day—and he's selling them to medical research."

When he returned home from his job at a pizza chain, Jack Plante found his wife in tears. She told him she would never forgive herself. Jack remembered the "crazy woman" who called, and now he turned away so that Jeannie would not see the tears in his own eyes.

The following day, Jeannie joined the callers trying to piece together fragments of conversations, descriptions, impressions. Several victims said the Mazda was light blue. Denise Tracey, who had given up her three-year-old Lab, Duke, was sure it was a hatchback. One victim recalled a young man fitting the Mike/Steve description who identified himself as "Rob." To the Neumans, an elderly couple, he was "Steve Jacobs," who had "Jewish blood." He promised them he would make a donation of $50 to the City of Hope Medical Research Center where Shirley Neuman was an outpatient.

Freida Marchese recalled that the young blond man who answered her ad had been stationed in Germany. He convinced the fifty-four-year-old woman to give up not only Eunice, the puppy she had advertised, but Eunice's mother, an Australian Dingo named Maxine. Mrs. Marchese, who was suffering from cancer, joined the telephone blitz. Within days, her hair turned completely gray.

Meanwhile, another name was added to the list. Steve/Mike told several victims that his fiancée was "Barbara."

At 6:15, Friday evening, January 21, Jeannie was wrapping up her calls. She had learned that half a dozen dogs and many more cats had been taken that day; she was feeling utterly helpless and emotionally drained. She had not slept a full night in nearly two weeks. One more call, and then she would go to bed. Jeannie dialed an 818 extension advertising "Three healthy, beautiful cats to a good home only."

"Was there a woman involved?" the cat owner asked. "Because there's a young woman here right now named Barbara and she's talking to my sister about our cats. She says she has a rodent problem, but we told her our cats are house pets. She's rather insistent that she wants them anyway."

Patricia Ossim described the woman. Her dark brown hair was shoulder-length, curly. She wore faded jeans and a hip-length beige leather jacket. She looked about thirty.

Pat listened carefully to the instructions Jeannie gave her.

"Barbara," Pat said, walking into the living room, "we would like some information from you to ensure our cats go to a good home, if you don't mind?"

Barbara smiled. "Sure."

"Do you have a driver's license?"

"It doesn't have my new address, I just moved," Barbara explained.

What kind of work did she do?

"Children's shows with animals," Barbara said. The name of that enterprise was Barbara's Cuddly Critters.

Where did she work?

"Out of my home."

Where did she perform these shows?

"Different places."

Did she have any references, friends, or family?

"No."

Barbara declined to give her last name or anything more specific than her needing mousers.

"We do want you to have our cats," Pat finally said as Barbara nervously prepared to leave. "But we would like to deliver them to your home to see where you live. Could you give us your address?"

Barbara hastily scrawled an address on a piece of paper. She said she would call later to confirm a time, and hurriedly left. The address was 15780 Cobalt, in Sylmar.

It might have been another phoney address, but the "detectives" took a gamble. The following morning, Bill Dyer located the modest ranch on a residential street in the Valley. Parked in front was a light blue Ford van. Sources at the police department told him that the vehicle was registered to Budget Boarding Kennel, on Norris Avenue, Sylmar. The owner of that kennel was a woman named Barbara Ruggiero.

Hours later, the USDA office in Sacramento confirmed the pet owners' worse fears. Barbara Ann Ruggiero and a Frederick John Spero had been issued a USDA B Dealer license #93-B-166, effective October 27, 1987.

By midafternoon, January 22, half a dozen victims and callers had installed a twenty-four-hour surveillance at Budget Boarding and Comfy Kennels. Two cars, two people in each car equipped with short-range police radios, were in place at all times.

Cars were also stationed near the Ruggiero house in Sylmar. In the backyard, large pens that resembled chicken coops had been sighted.

The cages were empty, but evidence remained of their inhabitants: piles of animal feces and maggot-ridden canned food.

The purpose of the surveillance was to find out exactly who was taking these pets and their destinations. Only then could the victims demand legal intervention based on documentation.

On the morning of January 25, Sheri Lamotta received a call from Christie McMicol, a legal secretary. The two had been working together for over three weeks tracking victims. Christie said she had an idea where Max might be. By noon, Sheri and a friend were scaling the chain-link fence at Budget Boarding.

"Nobody seemed to be there," Sheri later recalled. "The gates were locked. We didn't know if these people had pit bulls or guns, but I didn't care. My adrenaline was up." She later recalled stumbling over piles of debris, past a room stacked with cages filled with sick cats. "The place stunk, filthy. There was no water, little food. There were dead puppies in freezer bags on the counter. It was a horror."

A young woman* suddenly appeared in the hallway. She seemed to have just woken up. She was holding a boy of about seven. They were both shabbily dressed and filthy.

"Where the hell is my dog, you bitch!" Sheri screamed.

The woman said she was a boarder, that she had no idea what was going on, but she was going to call the police if Sheri did not leave.

"Go ahead, call the cops! I'll report you to Child Welfare and get your son taken away from you for living in this shit hole."

Sheri passed cage after cage of dogs barking wildly, jumping up on hind legs, tails wagging. And then she saw Maxwell. He was struggling past several large dogs who were crowding him out. Sheri flung open the kennel door and pulled him into her arms. Max's fur was matted with feces, he stunk of urine, but he was licking her face and Sheri was crying. Her companion grabbed a half-dead puppy and they fled under the protective eye of a surveillance team.

That same night, six surveillance vehicles were at their posts. After two days of little sleep, false alarms, and endless waiting, the mood was bleak. Dogs and cats had been arriving en masse at the kennels, but none had been taken out. Many of the "detectives" wanted to call in the police. Others argued that they first had to learn the animals'

---

*Terry and Cindi Phillips had left Barbara's employ a month before this new boarder had taken occupancy.

destination to make the theft case stick. The divisiveness was causing tension and ill will. Something had to give.

At ten o'clock a voice crackled over the police radio. *"Blue van moving out of Comfy. No animals."* A routine call. The van headed north on Norris about three-quarters of a block, turned onto Astoria, continued west, then made a left onto Bradley. Its driver suddenly jumped out of the van and raced up Tuxford.

"That's him—Steve/Mike," someone radioed.

Ralf was heading toward one of the surveillance cars. He thrust his head through the driver's open window. "What the hell are you doing!" he screamed. "What are you following me for!"

Maxine Lake, the driver, and Regina Eschelman, her partner, were too terrified to reply. Ralf raced back to his van and sped toward Comfy.

A dark-haired woman was waiting outside when the van plowed up the driveway.

The alert was sounded: *"Dogs being loaded at Comfy. Stay close."*

The blue van, driven by Barbara, pulled out, followed by Ralf in his Mazda. They headed west down Astoria toward San Fernando Road, closely followed by five surveillance cars. The caravan turned south on San Fernando and up Mission to the Golden State Freeway, heading north; two surveillance vehicles were lost in the maze. The radios went dead. As the cars wound on and off the freeway, two more vehicles fell back. Hours later the van would lose the last car, along a suburban street.

As the renegade van continued unfettered to its destination, at midnight, back outside Comfy Kennels—one block away from Budget Boarding—months of waiting were at an end.

Summoned by Bill Dyer and Sheri Lamotta, a dozen victims were milling outside Comfy. They were expecting the media, and Bill asked them all to wait. Max Neuman turned to his wife and he saw she was scared. She told him the darkness, the idea of breaking in, even to claim their own dogs, reminded her of the war. Her family had been taken from their homes in Poland by the Nazis on a night like this. Max held his wife's hand, and anger welled up inside him. He was a man who felt deeply about family and religion; Princess and Ace were like the hundred foster kids he had raised in the past twenty years. The anxiety had already damaged Shirley's frail health. And now they, respectable people who had never broken the law in their lives, were reduced to committing an illegal act.

Jack Plante told the crowd he was sick to death of waiting. He and his son, Montana, were going inside to find Butch. They scaled the seven-foot chain-link fence and Montana unlatched the gate. The Neumans followed and the others swarmed in behind them.

Shafts of yellow light from the flashlights some victims carried splintered the pitch-black darkness. The floors were wet, as if they had been recently hosed down. Still, the smell of feces and stale urine was unbearable. The barking was deafening.

Jack shone his light on each of the dogs. When they had passed the first row, he felt a sickening sensation in the pit of his stomach. He saw the same expression on his son's face.

Someone cried out, someone sobbed. It was all unreal, disconnected from life. The dangling chain of a light bulb brushed Jack's cheek. He reached up and switched on the bare bulb and the kennels were suddenly illuminated. Jack noticed a name scrawled on a chalk board: FLICKA. It was then that he saw a short leash dangling on a hook. There was no ID tag. But there were the same stains, the same marks, the same beat-up blue leather. He saw the tears running down Montana's cheek.

Jack Plante did not consider himself a violent man, but that moment he knew he could kill this Steve Johnson who had lied to his family and taken away big, good-natured Butch.

Several feet away, Mrs. Marchese shouted she had found Maxine. But where was the puppy? Deep in the kennels, Max Neuman called to his wife: he heard Ace and Princess barking. He'd know their bark anywhere, like he knew any of his children's voices. The stench nauseated Shirley, but she followed her husband. Alongside them, Vanessa Havens could not find Axel or Gypsy, her cats that had been adopted by a handsome young Marine who said his wife loved animals. Now, blinking back tears the pretty nineteen-year-old rummaged through the filth of the cat room. On a last hope, she flung open a closet door. Within there were stacks of cages, two cats huddled in each cage. The dim light reflected in their eyes and shone on the bronze tags that hung from plastic collars around their necks. One of the cats lay in a black and white heap; he looked dead. Vanessa cried out, unlatched the cage, and grabbed Axel. His body was wracked in spasms, his eyes coated with yellow mucus. A woman beside Vanessa was holding a small red collar with a bell and crying.

People raced out of the kennels with their dogs and cats that were covered with feces and stinking of urine. Ace and Princess, both emaciated, bolted out with Max and Shirley. Vanessa went home with only Axel. Many of the victims left with a collar or two. Most left empty-handed. Men and women were crying, only a few with joy.

• • •

On Tuesday morning, January 26, Lieutenant Bob Penia received a call notifying him of the prior night's events at the Sun Valley kennels. He could not believe what was happening. Sure, there had been rumors and a complaint had been filed about Budget: an actor thought his dogs had been "adopted" by Barbara Ruggiero for sale to medical research. Penia had sent one of his officers to Budget, but she had reported back that Barbara's partner, Rick Spero, told her, "No, we are not selling to medical research."

Now Penia hurriedly called his district supervisor.

Valley born and raised, Gary Olsen looked like Paul Newman, with his blue eyes and sandy colored hair. Like Penia, Olsen was a career humane officer. His manner was soft-spoken and unaffectedly courtly. In 1969, as a returning Vietnam veteran, he had applied for employment at both the LA Police Department and Animal Regulation. He was attracted by the security of a government job; he and his wife wanted to start a family. A position as an Animal Control officer was immediately available. Weeks later, when the LAPD offered Olsen a job, he turned it down. He was hooked on animal work.

At the time of Bob Penia's call, Olsen had served twenty years with Animal Reg. He had grown up in the agency as a protegé of its general manager, Robert Rush.

Olsen instructed Penia to dispatch an officer to the scene. He was concerned that the kennel might have been abandoned and that animals were being taken illegally from the premises. As to the allegation that Ruggiero was stealing pets for sale to medical research, Olsen told Penia, "No way." No kennels the Department licensed in the City of LA were allowed to sell to medical research.

At 3:30 that afternoon, Officer Ann Ivey was sent to Comfy Kennels. The conditions inside were appalling. People were milling about outside, screaming. The situation was tense. Officer Ivey secured the premises and reported back that everything was under control. Repeated calls were made to Budget Boarding, but there was no answer. No officer was dispatched to Budget, only a block from Comfy.

●　　●　　●

At about 5:00 p.m., Rick Sheker was waiting for his wife to return from work.

"I don't want to get your hopes up, babe," he told Candy as she opened the front door. He then explained what he had heard about the break-in at Comfy.

"That bitch!" Candy screamed. She raced to the kennel only 200 yards from her house and confronted the guard from Animal Reg. "Is there a black Lab inside?" She demanded to know.

The guard said he was not authorized to answer any questions.

"Please, just nod yes or no," Candy begged.

The officer said he could not respond.

Rick Sheker told his wife to stop crying and look at the man.

The officer was nodding. Yes.

●　　●　　●

At 6:00 p.m., Gary Olsen decided to check on Comfy on his way back home. He cruised by the kennel and saw it was secured by a guard. Then he headed up Tuxford to Budget. The scene there was quite different. There was a crowd of angry people shouting for the kennel owners to come out.

Barbara Ruggiero and Frederick Spero were at Budget Boarding when Olsen walked in. His first impression was that they appeared to be average, nice-looking, pleasant people who were clearly scared. For reasons they said they could not fathom, they were being harassed. Their lives were being threatened.

Olsen suggested they contact their lawyer and a meeting be arranged as soon as possible to determine exactly what was going on.

●　　●　　●

At 2:15 on January 28, 1988, Barbara Ruggiero and Rick Spero, accompanied by their then-attorney Hugh Seigman, sat in Olsen's office at the East Valley Shelter. Barbara had given Olsen a four-inch document that listed 141 dogs and cats she had acquired, most within the past four months. In many cases their prior owners had been listed; so had the destination of these animals.

At eleven o'clock the following morning, Olsen faced a crowd of victims, animal activists carrying signs, and reporters who had gathered outside the shelter for the Department's press conference. His job was to tell many of these people what had happened to their pets, and to try to salvage the image of the Department.

Robert Rush stood rigidly beside him. Olsen could feel his crackling tension.

Looking quintessentially Germanic with his manicured mustache and rigid stance, Rush recognized reporters from the *LA Times*, the *Valley News*, the *LA Herald Examiner*, and the *Daily News*, the same papers that had lambasted Animal Reg on countless occasions. These reporters would send today's announcement through the national wire services. This was a crisis situation, and Animal Reg had better come through.

Cautious, conservative, skeptical of change, the fifty-eight-year-old manager of Animal Reg badly needed a victory. An employee of the department since 1957, Rush was now in control of its $6-to-$10 million annual budget, and residents were questioning just what the Department did with all that money. Certainly there were no major efforts to spay and neuter dogs and cats, this despite chronic pet overpopulation. Rush was slow to make reforms, although he did support ending pound seizure and diligently implemented that legislation.

A consummate politician, Robert Rush recognized in the Ruggiero case a chance to redeem Animal Reg's shaky public image.

At 11:15 Gary Olsen announced his findings. Barbara Ruggiero and Rick Spero were operating, at the site of Budget Boarding, a facility known as Biosphere. Biosphere was a kennel licensed by the U.S. Department of Agriculture to sell animals to medical research.

Olsen then told the stunned audience that Ruggiero and Spero had acquired over 140 dogs and cats from the "For Free" ads: seventy-eight had already been sold to Los Angeles institutions. Cedars Sinai had purchased thirty-one; twenty-five of those had died in experiments. The Veterans Administration Hospital in Sepulveda had bought twenty-nine cats; twenty-one were dead. The eighteen dogs sold to Loma Linda were all dead.

By phoneying "adoption" papers, Ruggiero had also acquired thirty-nine pure-bred dogs from three LA county shelters.

The rumble of traffic plunged into the silent crowd. Nobody moved. Then all at once everyone reacted. Suddenly, everyone was shouting or crying. Jeannie Plante grabbed onto Sheri Lamotta for support. Sheri, with Max at her side, shook with sobs as news photographers converged on victims for close-ups and questions.

Sheri heard someone shout "We're going to Cedars!" and then the crowd piled into cars and vans and headed south on the Freeway to 8700 Beverly Boulevard, the LA campus of Cedars Sinai Medical Center, which had, for years, been the target of local activist demonstrations over its use of animals in research and alleged animal-care violations. By noon, victims and the press were converging on the glass doors of Cedars' Halper Research and Clinic Building. An activist attempted to gain entry but was rebuffed by a guard. "These people just want their pets," he screamed.

"Please, mister, I just want my dog," Jeannie Plante cried as tears ran down her cheeks. "Please just give me my dog."

A reporter from one of the LA papers who observed the scene would later recall, "It was heartbreaking to see those people who thought their animals were in good homes but were actually in research labs."

A fight broke out among activists and security guards. A paramedic unit was called to the scene and an injured man was taken away.

Cedars denied having any of the allegedly stolen dogs or cats. "Let these people bring proof," Ron Wise, Cedars Sinai's public relations director, told reporters. "This is just animal rights hysteria."

That evening, headlines in the Los Angeles press reflected the anger and grief of the victims who had been turned away. Many would later learn that their pets had died from invasive, painful procedures that included induced heart attacks, electric shock to the brain, and surgical instruction in classrooms.

The news of the pet scam shocked the insular City of Los Angeles. But similar headlines had become commonplace in communities across the country ravaged by pet racketeers: "Torture and Legalized Theft," *Minnesota Law Journal*; "SPCA Trying to Help Frantic Pet

Owners," *Burlington County Times*, New Jersey; "Fate of Dogs Given to Freetown Couple," *Standard Times*, Massachusetts; "Free to Good Homes May Invite Dog Dealers," *Times*, St. Louis, Missouri; "Stolen Pets Sold in Animal Slave Trade," *Northwest Arkansas Times*; "Warning Pet Owners: Dog Theft Increasing in Florida," *Putnam* (Florida) *Pennysaver*; "Dog Seller Finds Cop Nipping at Heels," *County Courier*, Vermont; "Be Alert Dog Owners," *Sentinel*, Georgia; "Lost Cat Traced to Quakertown Dealer," *Philadelphia Sunday News*; "Loving Pets Vanish," *Globe Magazine*, Boston; "Trucks Carrying Canines to Labs Uncovered," *Martinsburg Journal*, West Virginia; "Do You Know Where Your Pet Is?" *Tacoma News*, Washington.

Dog dealers like Ruggiero and Spero and bunchers like Ralf Jacobsen did not confine their sales to local institutions or even to those within their state borders. Via a well-organized, Mafia-style network, dealers stealing pets in Iowa were selling them to laboratories in Minnesota, just as pets stolen in Virginia were being sold to researchers in Pennsylvania.

Institutions were trying to buy further from home to avoid a local pet turning up in the mix. It was a lesson Cedars, Loma Linda, and the Veterans Hospital would quickly learn.

But LA was convenient for Barbara Ruggiero. It was home to her coterie of lovers whom she easily manipulated into doing her dirty work. The income was steady, the LA research industry anxious to patronize a local supplier. Greed was the singular motive for most of the 5000 USDA-licensed dealers and hundreds of thousands of unlicensed bunchers.

However, Barbara had not counted on the rage of the communities she pillaged or a formidable adversary 3000 miles away, someone who had spent decades tracking racketeers who were dealing stolen pets to research.

FIVE

Guerilla Warfare

"Pet theft is organized crime sanctioned by the U.S. government."

—*Mary Warner, Action 81*

October 1990
Northern Virginia

In autumn, the outbuildings and corrals of Mary Warner's cattle ranch are nestled in lush, mustard-green hills that take on an amber cast at twilight, the kind of bucolic landscapes glorified by the nineteenth-century English painter Turner. The Shenandoah River courses calm and crystalline through woodlands dappled in auburn and gold.

This is a region of the country where Old Money raises its thoroughbreds, rides the Hunt and "Point-to-Points," and introduces its daughters at cotillion balls. It is a setting that befits a distinguished ancestor of one of America's founding families.

But when Mary Warner looked out her window she did not see a privileged landscape; she saw a battlefield.

Mary's office in two rooms of a small outbuilding she called the Dog House was cluttered with documents and maps; the walls were densely papered with photographs and posters, most handwritten, with the words "REWARD, NO QUESTIONS ASKED" in urgent letters.

The first thing that caught the eye was a large map of the U.S. which was virtually covered with tiny black circles. The circles traced routes down the East Coast to the Sun Belt, through the heartland of America, along the Mexican border, and west to California, Oregon, Washington, and Alaska. Beside that map, Mary kept a running list— "updated casualties," she called the figures obtained from police, humane officers, sheriffs, veterinarians, and victims tracking the national dealer network in every state in the country: 1000 dogs and cats missing in one month in Indianapolis; three hundred dogs and cats missing within one week in Billings, Montana; five hundred cats and dogs missing in six months in Honesdale, Pennsylvania. There were pencil scribblings of recent reports: eight-six dogs vanished in one month in Hagerstown, Pennsylvania; twenty big dogs disappeared from one neighborhood in Bristol, Tennessee, within a week. And the number that topped them all: 10,000 dogs reported missing in Rochester, New York, within six months.

"I know all these animals," Mary said, and the tears she fought back made her hazel eyes shine. "I've spoken with so many of their owners. I've heard grown men cry on the phone and the same stories over and over again: trucks and vans cruising neighborhoods picking up dogs—Standard Poodles, English Setters, Golden Retrievers. Hundreds of dogs gone within days. I've gotten reports from Pennsylvania where kids in one area earn $2 just to tell dealers where the good dogs are kept outside."

Mary was a diminutive woman with large bright eyes and curly gray hair. The image of this octogenarian, a grandmother of twelve, fighting in the trenches seemed surprisingly conceivable. Her will alone could sustain her. "So many people are still looking for their pets, after years and years when there is really no hope at all, but they need to hold on to something."

Mary knew the power of faith. She had heard the same stories from pet owners again and again, but her belief that someday, somehow, this criminal network would be exposed helped her endure. Now at age eighty, she has taken too many of these painful journeys with hundreds of thousands of people she has never met. Their testimonies filled her office—reports of cult activity, sightings of skinned dog and cat carcasses for the fur trade ("genuine wolf collars"), bins of puppies at dog auctions that would be used as live bait to train fighting dogs. But by far the most valued, steadiest client for these stolen dogs and cats was the animal research industry which had at its disposal billions of taxpayer dollars and can pay a premium for preferred research subjects: family pets, no questions asked.

The accounts of grieving owners provided the first glimpse into the nationwide pet-theft business, a criminal network licensed by the government to serve the market demands of the medical research industry.

It all started for Mary shortly after a *Life* Magazine cover story, "Concentration Camp for Dogs," shocked an unsuspecting public in February 1966. In a stunning photo essay, *Life* exposed a pet-theft racket that was then supplying the nation's laboratories with "two million dogs and cats each year."

Mary recalled: "People got a glimmer of what was going on in 1966. But no one really knew the incredible scope, who was behind it all."

Neither did she. At that time, it was far from Mary Warner's imagination that she would be a principal front-line fighter battling "organized crime sanctioned by the USDA and financed by the medical research community."

"Concentration Camp for Dogs" was prompted by the discovery in a New York laboratory of a family's Irish Setter that had been reported missing in Pennsylvania. The dog was the beloved companion of a terminally ill child. The February 4, 1966, issue of *Life* displayed horrifying photos of emaciated dogs staked out at dealer kennels that were nothing more than junk yards; dogs crammed into crates along with pigeons and cats; carcasses of animals strewn in backyards amid the skeletons of auto parts. A Collie which the Humane Society could not rescue for lack of space in its vehicle was "too weak to crawl to the frozen entrails" that were his food.

Hours later, the Collie itself was a frozen carcass.

Life reported that these animals were "family pets, trained to obedience and easy to handle . . . stolen (and sold) to dealers. Some dealers keep big inventories of dogs in unspeakably filthy compounds that seem scarcely less appalling than the concentration camps of World War II." Auction prices in 1966: dogs, 30 cents a pound; puppies, a dime.

One reader actually found his own dog through that article. *Life's* photograph of a dog at the Harvard Medical School turned out to be his "missing" pet.

The research industry had, by that date, a one-hundred-year history of taking dogs off the streets to fill its laboratories. Franklin Loew, Dean of the Tufts Veterinary School, would explain, years later, that the rationale behind using dogs in the first place was "because they were just sort of running around. They were available."

Citizens were outraged and Congress was forced to act. At congressional hearings over a proposed bill to address this scandal, law enforcers across the country, Humane officers, victims, animal protection groups, and congressional delegates representing victimized constituents presented damning evidence of "the shocking abuse of the use of stolen animals for medical research." They told of dog farms that were "collection points for stolen dogs," of deplorable conditions at those kennels where as many as half the dogs died before being shipped for experimentation. Midwest officials spoke graphically of auctions where researchers purchased dogs and cats:

> Fifty to sixty dogs in pens and cages, some so sick with distemper they could not get up. They were moaning, frothing at the mouth, and shaking violently near convulsion. Puppies with feet so red and infected they could not even stand on them. A beagle so sick it rocked from side to side with pain. A cat so ill its head was swollen almost the size of its body.

A chief investigator for the Maryland State Police described one lab supplier's kennel in Maryland as "a tangle of wrecked automobiles, trucks, body parts (housing) over one-hundred dogs. Dogs were desperately licking at frozen water pans attempting to drink. There were dogs scratching and clawing at frozen pieces of bovine entrails, their only food. It did not take a veterinarian to determine that many dogs were emaciated and starving."

Few speakers were more imposing than the stately woman who had founded the Animal Welfare Institute. Christine Stevens, president of the Washington, D.C. animal protection organization, strode to the podium with a letter from a victim of pet theft. A stunning fair-haired socialite—her husband had been Treasurer of the Democratic Party and founded the Kennedy Center for the Arts—Stevens told Congress how one family's dog, Lancer, had escaped from Harvard Medical School during transport, circumstances confirmed by the school itself. Lancer, thin and voiceless, had miraculously made the journey home to his family of seven.

Stevens told a stilled audience that Lancer's owner wanted to know: "Why do they keep these places a secret? Why isn't the public informed of these places so they can look for their lost animals there?"

Dr. Bernard Trum, director of the Animal Research Center, Harvard Medical School, conceded at that hearing: "We often have the pleasure of returning strays to their owners."

New York Congressman Joseph Resnick, the bill's sponsor, read aloud a letter he had written to the Secretary of Health, Education and Welfare, the Honorable John W. Gardner, which summarized the critical situation:

> The animal procurement practices of the Nation's research laboratories have become a national disgrace. Family pets, dogs and cats are stolen off the streets, sold to disreputable dealers and eventually wind up in the hands of suppliers to hospitals and laboratories. Most users of these animals are indifferent to the manner in which their suppliers obtain them.

Resnick explained that another undesirable side effect of such a procurement system is the poor health of the animals. "Underfed, sick, and weak (they are) hardly ideal specimens for meaningful constructive research. I have been told by doctors of many laboratories that they frequently buy two or three times as many animals as they actually need because of their condition."

Resnick could not have known then that the same conditions would still exist thirty years later.

During those hearings, the research industry exercised arguments it would use for decades to come. The industry railed against the Animal Welfare Act and its implication that researchers were "uncaring monsters." It vehemently resented the notion that it had to be subject to outside monitoring from the government, and, in response, brought out for battle its heavy artillery: representatives of the American College of Pharmacy, the American Heart Association, the American Association for the Accreditation of Laboratory Animal Care, the American Dental Association, highly placed researchers from the University of Pennsylvania, NYU Medical Center, Montefiore Hospital, Veterans Administration hospitals, the University of Chicago, Harvard; even dog dealers. The same entourage would lobby during the next two decades to try to overturn the humane provisions of the Animal Welfare Act; only the faces of those testifying would change.

The American Medical Association lauded researchers' high standards and decried the allegation that its members were buying stolen pets. The consensus among the industry was that the Animal Welfare Act was "covertly punitive."

But what exactly was so offensive about humane conditions, better care, proper record keeping, proper procurement, proper identification of animals, and inspections of dealers and research facilities? The research industry declared that these measures were "purposeless, unnecessary, costly, and impairing the efficiency of medical research."

The "medical progress" argument would provide a resilient and compelling smokescreen, diverting public and legal attention from the real issue: criminal pet theft.

Meanwhile, the National Institutes of Health, which funded animal-research experiments with billions of tax dollars, was pressuring President Lyndon Johnson to kill the bill. NIH managed to exempt government agencies—including the military, which ranks among the largest users of animals for experimentation (500,000 animals each year)—from the regulations. Government institutions were not required to report the numbers of animals they used or how they used them, or even to follow the humane standards of the Animal Welfare Act. Thus, hundreds of thousands of stolen dogs and cats slipped through a government loophole created by industry lobbyists.

Wringing its hands at the hearings was the Animal Welfare Act's future custodian, the U.S. Department of Agriculture. USDA officials blatantly tried to devise ways to wangle out of enforcing the legislation. Secretary of Agriculture Orville Freeman argued, for one, that USDA was not "directly concerned" with the issues. When it was clear USDA would be saddled with the Act's enforcement, Freeman asked Congress for a "broader definition of supplier" to allow researchers greater leeway in purchasing dogs and cats. His suggestion ran counter to the very purpose of the bill.

The resulting Animal Welfare Act was a relatively short document that established minimum federal guidelines and standards for the procurement, care, and handling of dogs and cats used in laboratories.

Dog dealers would have to be licensed and inspected by the USDA and would be required to comply with certain standards of care, as would medical research facilities. Selling animals to research without a license would be punishable by stiff fines. It would be unlawful for researchers to buy from an unlicensed dealer.

Nonetheless, lobbyists for the medical research and animal-use industries succeeded in trimming the Act's power. The tens of millions of rodents used each year in laboratories were omitted from protection under the Act.* Also excluded were animals in exhibitions: zoos,

---

*In 1992, a federal court ruled that rats and mice should be included under the Animal Welfare Act, but USDA is appealing that decision.

circuses, rodeos, wrestling and marine mammal shows. The Act did not protect horses in the multimillion-dollar racing industry, the millions of animals in shelters, and the over 100 million dogs and cats owned by individuals. It did not protect birds, reptiles, fish, amphibians, invertebrates, or the billions of farm animals consumed each year. And the Act made no legal provision for the protection of animals used in painful research. Institutions were, quite simply, granted autonomy in the way they used animals in experiments. To this day, federal law and professional guidelines set by the National Institutes of Health still expressly permit the witholding of anesthesia, analgesics, and tranquilizers during painful animal experiments.

The *Life* article had a special significance for Mary Warner. In 1966, the Warners had sold their business in Minneapolis and moved to a farm they had purchased in Clarke County, Virginia. "When I read the *Life* article back in 1966, I was horrified to learn that we were moving into the center of pet theft," Mary recalled. "Clarke and Fredericks counties had very bad larceny rings. Even the dog warden in one of the towns was known to be stealing dogs and cats and selling them to research. People here were familiar with the terms 'dealer,' 'buncher.' Some of my neighbors had placed signs out on their lawns: 'Dog thieves beware. We know you are coming.' It was quite terrifying."

•   •   •

On one warm autumn afternoon, the door to Mary's office was open, allowing access to whoever needed a bit of reassurance, comfort. Trubie hobbled in, his small, dappled body muscular from the effort of walking on only three legs.

Trubie had lost his front leg in a car accident. Wounded and needy he was adopted by Mary. He was one of many three-legged, half blind, traumatized dogs and cats rescued from auctions or from alongside interstates and back roads, where they had been dumped by buncher trucks, or shot and left for dead. They made their home at the Warner farm, along with the horses Mary still rode and jumped, the cattle, and the visiting children and grandchildren.

The Warner's white colonial perched on a hilltop overlooking the platinum Shenandoah. Within, the home was sunlit, cozy, welcoming, with lush bright red (Mary's favorite color) carpeting, overstuffed sofas and armchairs in chintz, dozens of redwood-framed photographs of the Warner family on horseback, at the Hunt clubs, showing, jumping.

Mary drove a 1974 convertible VW bug—red, of course. She kept Christmas decorations up, year-round: "It's so cheerful and makes things so much easier when the holidays come around."

The Warner homestead teemed with life—new, battered, hopeful.

In 1967, the first year USDA began licensing dealers, several of Mary's Beagles vanished from her farm. Then her Bloodhound, Hoover, disappeared. Everyone knew Hoover, but he never strayed beyond the perimeters of the house. Mary searched for weeks, notified the local ASPCA and all the sheriffs in the tri-county area. Then a call came from Leesburg. Hoover had been found near the National Institutes of Health, two hours from Mary's home, clear across the Shenandoah River. At that time the government research facility maintained its dogs in stacked cages outside, to the rear of the building. A resident living nearby saw the dog wandering collarless, and he called the ASPCA. They told him no one had reported missing a Bloodhound. He then called several sheriff's offices, and Mary was reunited with Hoover.

That same year, millions of "Hoovers" were ending up at dealers and in laboratories.

When it came time for the USDA to enforce the Animal Welfare Act, its resentment of the task became clear. One of the first dealers to obtain a USDA Class B Dealer license allowing the purchase and sale to research of "random-source dogs and cats"—that is, those not specifically bred for research—was Ervin Stebane. For years, the cantankerous owner of Circle S Ranch in Appleton, Wisconsin, had been the chief supplier of dogs and cats to the University of Wisconsin, despite conditions at his kennels which were described by the *Twin City News* in a 1960 article entitled "Shocking Filthy, Frozen Death at Nearby Dog Farm":

> Recently warm sunshine melted away the snow which covered at least two piles of dead puppies. . . . Children of a tenant farmer living on the site told of the dog dealer destroying 'sluggish' puppies by beating them to death against the barn walls while the children watched.
>
> . . . Inside (a dog coop) were three half-frozen calves' carcasses on which the dogs were gnawing and fighting. . . . Very bad condi-

tions and violations of the state sanitation code (reported) by an official of the State Department of Public Health.

In fact, the condition of Stebane's kennels had been cited in the congressional hearings to demonstrate the desperate need for an Animal Welfare Act. Yet, Stebane—with his piles of dead dogs and his cattle trucks hauling dogs in twelve inches of their own feces—was approved for a license by USDA inspectors.

Stebane's USDA license (#35B9) earned him a substantial income over the next twenty-six years, as well as the loyalty of the University of Wisconsin, Appleton Medical College, Fromm Laboratories, Madison Area Technical College, and USDA, which stood by him as media, former employees, and eyewitnesses documented theft of residents' pets and atrocities at Circle S Ranch.

BAR WAN Rabbitry and Kennels was another of USDA's first licensees. The Crocker, Missouri, kennel was then a mom-and-pop backyard business. The conditions under which BAR WAN was licensed were described in a report by the USDA inspector who visited in 1967: two-hundred adult dogs were tied without shelter, many without water. Chicken-wire hutches contained a passel of puppies, many sick, the water in the container too high for many of them to reach. A crate of cats was so crammed that the inspector could not count them.

BAR WAN would become a multimillion-dollar facility complete with plane transport service guaranteeing delivery within twenty-four hours to any institution in the U.S.

USDA also awarded one of its early licenses to Sam Esposito, owner of Quaker Farm Kennels and patriarch of what would become known as the Pennsylvania Dog Mafia.

Mary Warner would set out after Sam Esposito shortly after her beloved Copper was stolen.

Mary's office phone rang continuously—press, victims, animal protection and consumer organizations, law enforcers. Often she let the answering machine pick up and the announcement take over: ". . . *Pet theft is extremely prevalent now across the country. Do not leave your dog or cat alone where it may be seen or taken.*"

It was a very personal message.

The bronze plaque prominently displayed on Mary's desk was a humanitarian award forged in the image of Copper, her German

Shepherd. In 1974, Copper was stolen. Mary had seen the suspicious van at the edge of her property, the same van her neighbors observed the day their own Labradors and Shepherds disappeared. The loss of Copper still brought tears to Mary's eyes.

In her search for Copper, she started networking locally, spending hours on the phone. She began to hear similar stories of human tampering: a cut rope or leash, an unhinged yard latch, a collar tossed on the driveway, an unmarked truck seen picking up dogs, pets taken from backyard kennels, from front lawns and neighborhood streets. Quantities of pets missing in short periods of time, often the same breed of dog.

A year after Copper's theft, Mary was asked to testify before Congress on behalf of strengthening the Animal Welfare Act. "I was just an ordinary person and I was so hopeful, so confident in the democratic system of government where people like myself were listened to. Over the years it has become clearer to me that pet theft can be stopped—and prevented—by ordinary people."

Mary explained to Congress how it all worked. "Bunchers," the low men on the totem pole, were the "foot soldiers" of the dealers, who received their orders from the research institutions: 550 male Labs needed in three weeks; 126 German Shorthaired Pointers in six weeks; 300 neutered, shorthaired, big-chested dogs in ten days. Quantity, type, due date. This is an urgent supply-and-demand business.

Bunchers are an inventive lot. Cruising neighborhoods, these men and women devise various techniques for culling dogs and cats: stun guns, a bitch in heat, high-pitched whistles. In Matthews, Virginia, in 1984, bunchers even posed as Animal Control officers and went from house to house, duping residents into giving up their cats which supposedly had a contagious virus.

But there was an even more insidious means of obtaining prime research subjects: the "Free to a Good Home" ads in every newspaper in every town in the county. "Free to a Good Home Only" was a gold mine for specific breeds and types most often requested by research and not always found in quantity at pounds: Labradors, Shepherds, Huskies, Golden Retrievers, and their cross breeds.

Color photos hanging on the walls of "The Dog House" showed Retrievers, Shepherds, and Huskies posed in backyards, on family trips, playing with children. They reminded Mary of the millions of dogs and cats that never came home. Two million of the five million pets reported missing each year have been victims of pet theft. Mary explained, "The demand is extraordinary considering the billions of dollars in grants and incomes for researchers using animals. And there are certain kinds of dogs researchers prefer. Where did the researchers from the lab expect to get one-hundred and twenty-six German Shorthaired Pointers? Well, residents near the lab soon found out."

•   •   •

It was nearing the dinner hour and Leon Warner poked his head into the office. He was tall, gallant, protective. "Don't tire yourself, Mame," he said affectionately. He had nicknamed his wife after the movie *Auntie Mame*, about an exuberant, indomitable lady. Their mutual adoration had been nourished over sixty years.

"I don't know what I would have done without him," Mary said. "Leon has been a pillar of strength for me."

They met at Smith College in 1931. Mary Case, as she was then known, would graduate summa cum laude, a Phi Beta Kappa with a degree in history; her dream was to become a concert pianist. After graduation she left for New York City and a rigorous tutelage under a Polish countess whose grandmother had been a pupil of Chopin. Mary envisioned an artistic life dedicated to her three loves: music, horses, and dogs.

Leon Warner was a dashingly handsome Dartmouth graduate, also from a gentrified Minneapolis family. He was three years her senior and on his way to Smith for a date with another student. He and the attractive, spirited freshman bumped into each other in the hallway of her dorm. It was love at first sight.

The Warners and the Cases were wealthy, opinionated, hard-working clans who traced their roots to the English settlers. One of Mary's ancestors on her mother's side was Thomas Janney, a Quaker who arrived with William Penn in 1683, seeking freedom to worship; Janney settled in Loudin County, Virginia. His wife, Hannah, led the Quaker movement against slavery in the late 1700s, and during the 1800s the Janneys were part of the Underground Railroad, helping slaves escape to freedom. "It's the same blood flowing through our veins, fighting against the trafficking in human beings and, now, in animals," Mary said proudly.

The Case men, meanwhile, pioneered some of the earliest efforts for racial equality and unionization. "Our families were never followers. We were determined to see justice done."

Leon shared Mary's love of music—he was a cousin to Sir Edward Elgar, the British composer best known for the work "Pomp and Circumstance"—and he too was an avid rider. Dapper and energetic, the eighty-three-year-old gentleman farmer worked the same hours as his young ranch hands. The work ethic and physical stamina were in the Warner and Case blood. While the Cases were battling on the political front, the Warners were breaking records at the earliest Olympics trials and founding companies.

"I married a dangerous woman," Leon said smiling. He quietly shut the door behind him and called to the dogs for their evening walk.

"All my life I've been a loner," Mary explained. "I have always fought out against the status quo, against things that are not fair. But having that kind of companionship has been a solace when I felt I could not endure yet another heartbreaking story."

Leon later confided that he has lived for over twenty years in fear for his wife's life.

After dinner, Mary returned to her office to listen to her messages. Sue, one of her spotters in Ohio, had called. Sue worked as a beautician during the day and on an assembly line at night. "You're not going to believe what's goin' on here, Mary: 4000 dogs at the Kenton auction, and we got license plates of the research trucks buyin'. Finally! And we got a lead on the pet scam wire fraud. A trucker's answering "Lost" ads that offer rewards and he's callin' these people, tellin' them he's got their dog and to wire him money through Western Union. The FBI's busted him in Texas, but we can't get action here." The message cut off, but Sue called back. "He's milkin' people in Ohio and Indiana and now Connecticut, makin' a fortune. And we got three kennels dealin' dirty dogs. It's bigger than we thought."

Cale in Texas asked for a law enforcer referral in Oklahoma: "We got the shipment days from that Arkansas dealer to the University. We need someone to bust his driver at the border."

A victim in Reading, Pennsylvania, called to report that her neighbors' Labs and Shepherds were disappearing; where should they look for them?

Other calls were from reporters in Pennsylvania, Wisconsin, Arkansas. There was a Linda Elliot referral.

The godmother of the front-line fighters jotted down the messages. She would spend the rest of the evening and early the next morning returning calls, cross-referencing information, connecting front-line fighters across the country.

The pet-theft tracking network was working. So was the dealer/researcher network.

Nowhere is the scope of this network more visible than at dog auctions in the Midwest, where dealer and buncher trucks from virtually every state in the Union converge to swap stolen dogs and cats, switch license plates and carloads at the borders, phoney-up paperwork, and fill orders from virtually all the universities, hospitals, and products-testing companies, even military bases that conduct research using dogs and cats. Many of these institutions send their own trucks to the auctions.

The passport into these nationwide syndicates is a license issued by the United States Department of Agriculture.

SIX

Family Business

"Hell, this is our business. We're federally inspected.
We're legit. We don't have nothin' to hide."
—*Danny Schachtele (Class B USDA License #43B032)*

June 1990
Rutledge, Missouri

As it reaches Rutledge, Missouri, the Middle Fablus River, a brown snake of a waterway that feeds on the Mississippi, tunnels under the junction of Routes V and M. It dampens a landscape pocked with telephone poles, pig feeders, and fields of ragweed, then veers west toward what was once Pony Express territory, leaving Route M to track its course through downtown Rutledge.

Here, Route M becomes Main Street, a desolate road framed by boarded-up store fronts, deserted houses, two grocery stores, a fire department, a white-steepled Baptist church, and a brick post office.

The ghosts of cattlemen who founded this town in the late 1880s, when the Sante Fe Railroad reeled in northeast Missouri, hover among the rusted, broken rails. Except for an occasional Chevy pickup with a .22 slung on its rear window rack and a pack of Coon Hounds in its bed, there is, most days, rarely any sign of life in the heart of Rutledge, population 138.

In June, what was planted in April had taken root on the small farms within a few-mile radius of the town. Corn, grown mainly for livestock, was six inches high. Shoots of grain sorghum, a cash crop popularly known as milo, were rearing sturdy heads. Soybean plants bristled with green leaves. Masses of wild blackberry bushes were heavy with fruit.

But it might just as well have been winter, when temperatures plummet to five degrees and snow banks upward of three feet lie along the curb by the post office. The same sense of isolation pervades. It is a state of mind that translates into the hard features of the citizens of Rutledge. It is the kind of pinched loneliness that deepens under the relentless sky.

If the traveler were to recall the drive to Rutledge, his memory would be of flat sky, scattered yellow neon motel signs flashing "Vacant," and the tips of six-foot ditchweed—seemingly disembodied from their roots—wavering in the darkness.

This is the heart of dog-dealing country. These asphalt highways and dirt roads are conduits for an underground network that transports millions of dogs and cats stolen from every state in the Union for sale to medical research. It is a network not often visible, except on auction weekends at pin-dot towns like Rutledge, Poplar Bluff, Boonville, Salisbury, Iberia, and Vichy.

Since dawn on this second weekend in June, caravans of trucks and campers had been making their way to the northeast corner of Missouri. Some of the roads were slick and pitted with mudholes from the previous night's rain. The electrical storm had done little to relieve the humidity of the 105-degree week. It was going to be another sticky scorcher.

By seven o'clock, vehicles were plowing past the weatherbeaten sign that announced their arrival at Irvin Johnston's Flea Market Hillbilly Gun and Dog Auction, Rutledge, Missouri: pickups with campers, station wagons, and U-Hauls lugging cages, some empty, some filled with two or three dogs a cage. License plates from Missouri, Arkansas, Illinois, Oklahoma, Tennessee, Minnesota, Iowa.

At the height of the weekend, as many as 8000 visitors would barter for everything from homemade jelly and leg-hold traps to medicinal fox urine and AK-47s. Mostly, though, people came here to deal in dogs—as many as 3600 dogs on a good weekend. About 85,000 dogs, most of undocumented origin, are sold each year through Missouri dog auctions, predominantly to research buyers.

The flea market vendors parked in a muddy lot gouged with tread marks and started unloading and stocking booths. Transports lugging dog boxes switched into four-wheel drive and barreled toward a clearings in the woods behind the market where the concession stand was located. They set up near the gun booths, under signs saying "No Gambling," "No Drinking," and "No Boisterous Conduct Allowed."

Dog dealers—nearly a dozen by 7:30 A.M.—were sweating in torn tee shirts, wiping sweat from under feed caps. Some were bare-chested in muddied overalls, bellies slung over their work pants like slabs of white pork. They pulled out by a leg, head, or tail whatever has survived a journey of hundreds, oftentimes thousands of miles without food or water in wire cages and crates stacked atop pickups or crammed into school buses whose innards had been stripped.

The dogs that had been picked up en route to the auction still had shiny coats, bright eyes. The ones that had been penned up for days and hauled across state lines looked shell-shocked.

The "serum dogs," so named because of the popular belief that a serum made from the blood of dogs cures cancer, were hitched together chain-gang style between trees and the fenders of trucks slung low in the mud: mixed-breed Shepherds, Hounds, Beagles, Terriers, an occasional "retired" Greyhound.

A Beagle mutt gnawed at his chain. The jagged metal slit his jaw; bloody saliva coated the metal.

"Hey, cut it out!" A local dealer whammed the dog with his boot. There was a sickening crackling sound as the animal collapsed in his own vomit.

Puppy-mill buyers looking for breeding bitches or studs to produce the 400,000 puppies sold each year to pet stores, and dog fighters needing live bait to "blood" fighters, came regularly to Rutledge. But the medical research buyers were the mainstay of the business, had been for over forty years.

The quality of dogs is denoted by their category. "Serum dogs" carry the longest guarantee: seven to ten days alive after purchase. If the dog died of "natural causes" before its guarantee expired, the laboratory purchaser could technically demand a refund or substitute from its dealer. Few, if any, do. High mortality rates, as high as 50 percent, of random-source dogs are simply factored into the cost equation when researchers apply for grant money.

"Junk dogs," often sold "by the pound" and bought by product testing companies like US Surgical, are guaranteed alive no more than one week. "Acute dogs" carry the shortest guarantee, forty-eight hours alive after purchase and therefore are used quickly; some, for instance, may be embalmed by biological houses for classroom dissection sale. The small dogs would be sold to puppy millers; toys and puppies, were sold as live bait to train fighting dogs.

Whatever remained by Sunday night would be dumped among the usual road kills that littered Routes V and M after auction weekend.

Kids wearing skimpy, overwashed clothes played amid the gun booths and dog cabs. Some hung around the concession stand, manned today by PTA volunteers furiously cooking bins of ham hocks and navy beans, and hoping to do as well as the Lion's Club, 4-H, and the fire department which have earned up to $4000 on prior auction weekends.

There was a rush of oven-hot air as the doors of a serum truck swung open. "Forty-five to eighty pounds," the research buyer called out. He was offered an armful of squirming black pups.

"No bait."

"Bait hell, them's serum dogs. Them types been keepin' me in steak for a month." The hefty dealer threw the puppies into a bin where "bait" squirmed. He offered up an adult black Labrador. "$250," he said.

There was an order out for big-chested dogs, and this healthy, young Lab was being offered cheap.

The dog was lifted to the loading platform, then shoved into a filthy wire cage where a tan and brown Shepherd was cowering. Blood coated the loading platform.

It was a humid 92 degrees and only eight in the morning. The smell of fried possum and turkey nuts* sizzling at campsites mingled with the stench of dog feces and the acrid smell of sweat.

A Shepherd mix was shoved head first into a wire cage and hoisted into a white research van. A shaggy Benji-type with wild, bloodshot eyes was pulled from an airline carrying cage, grabbed by the neck, and tossed into the truck. Several $20 bills passed hands.

"This is a family business," said USDA-licensed dealer, auction-goer Danny Schachtele. "We're just country people makin a livin'. That's all we're doin'." Schachtele supplied dogs, at about $250 a head, to the University of Illinois; Alton Oschner Medical Foundation, Louisiana; University of Oklahoma Health Science Center; Oklahoma City VA Hospital; Kirksville College of Osteopathic Medicine, Missouri; and others.

"Where there're dogs, research trucks are gonna be there," auction

*Testicles.

owner Irvin Johnston explained philosophically. "Some weekends I have as many as 3600 dogs. They come from practically every state in the Union. I even got a truckload from Alaska. Research trucks come here. They get orders. Mostly they want dogs thirty-five to fifty pounds. One fella used to come and take them all the way to Florida. The reason I knowed was he wouldn't want but large dogs. Had to weigh a hundred pounds or more. He said he got an order from a certain hospital in Florida."

For forty-five years, Irvin Johnston had been running the auction on his northeast Missouri farm. A robust man in his late seventies with a shock of white hair and twinkling Santa Claus blue eyes, Johnston had been mayor of Rutledge for twenty-three years; he continued to be its economic patron, as well as local poet and a country songwriter. Over 20,000 of Johnston's business cards circulated each year throughout the Midwest bear his original anecdotes and philosophy, including an ironic one: "Since man is man's worst enemy, and dog is man's best friend, the sooner the world goes to the dogs, the better off man will be."

"I know one dealer may have orders in a half a dozen different places," Johnston explained. The auction block was reserved for trading Coon and Deer Hounds, Beagle Hounds, Fox Hounds, Running Hounds, while dealers bought "on the ground." "I got a call from Chicago here the other day, the Vet School, and he wanted to know if I could provide them with forty, fifty, a hundred Beagles. That's one good thing about here. It gets a lot of dogs out of the country that would otherwise be loose and killing people's livestock and getting into the dog pounds. They're worthless in that respect. Those dogs dumped along the road would be better off going to the research end because they would go ahead and take better care of them."

According to on-site investigations by humane organizations, eyewitness reports, sting operations, government "whistleblowers," local and national reporters, and TV crews filming undercover at various auctions including Rutledge, there was little or no verifiable documentation of the source of the 85,000 dogs that passed through Missouri auctions each year. Without documentation, there was no way to determine to whom these dogs once belonged. Dog dealers interviewed by the *Kansas City Star* in a July 1990 series on dog theft and auctions readily admitted to lying on USDA record forms about the source of their dogs.

The *Star* series also cited one USDA official's estimate that nearly two dozen of the largest "random-source" dealers in Missouri and Kansas annually grossed more than $100,000 a year. According to insiders, that figure was vastly underestimated.

The fate of these auction animals was more easily assessed. Some would be burned to death in experiments conducted by the U.S. Army to determine the effect of radiation in the event of World War III. Others would be force-fed toxic household cleaners in order to protect companies from product liability.

U.S. Surgical Corporation, the country's leading manufacturer of surgical staplers and a user of dogs in sales training, was a steady customer of one of Rutledge's regulars: Sam Esposito, the owner of multimillion-dollar Quaker Farm Kennels (USDA license #23BJ) Esposito reportedly bought "junk dogs" sold "by the pound" for $10 to $20 a piece and hauled them back east.

For decades, Esposito had routinely sent his forty-foot hob trailer and fleet of pickups from his Pennsylvania kennels to buy dogs and cats at Rutledge and at Kentucky, Illinois, and Indiana auctions. According to an internal delivery list, his clients included Yale University, Sloan Kettering Cancer Institute, Johns Hopkins, Princeton, Mount Sinai School of Medicine, and the National Institutes of Health, which told the public it bought only "purpose bred" dogs and cats raised specifically for research.

According to a Virginia state police investigation, Esposito also bought from "people ripping dogs off the street."

Over the past two decades, Esposito had been found guilty of violating state and federal cruelty, transportation, and health laws and of failing to provide accurate records and identification of the animals he acquired. But none of these violations had put a damper on his income, which averaged nearly $1 million a year. Esposito, a short, grandfatherly man now deceased, had friends in high places.

"Sam owned the East Coast," said Kathy Schweitzer, a longtime Pennsylvania Humane Officer who often spoke with Esposito. Their conversations prompted her to coin the phrase "Dog Mafia," referring to the powerful interrelationships among dealers, law enforcers and researchers across the country.

"Sam was a regular down here," Irvin Johnston confirmed.

On this steamy morning, several of Mary Warner's Midwest contacts, wearing dirty old clothes, and newspaper and TV reporters

dressed as shabbily, mingled unnoticed among the dog and gun dealers. For the past several weeks, pet-theft hotlines across the Midwest
had been buzzing with reports of hundreds of missing Shepherds,
Labs, and Huskies. One Missouri organization, Protect Our Pets, had
been receiving two hundred calls a week for the past three weeks
from frantic owners reporting their big dogs had vanished without a
trace. In Arkansas, COMBAT, another pet protection group, had also
reported high numbers of large dogs missing. And there were other
telltale signs of theft: the "Pets Lost" sections of local newspapers
indicated large breeds were missing.

Entire neighborhoods from Clinton County, Missouri, to Stark
County, Arkansas, were being picked clean.

Those kinds of numbers and breeds meant only one thing: the call
was out from cardiovascular researchers needing broad-chested dogs
for experiments. Odds were, dogs of that quality were being sold
direct to the laboratories, but there was a chance they were being
pipelined through the auctions. The pet-theft patrollers, including a
housewife and a teacher, jotted down license plate numbers of pickups with dog cabs. Reporters wandered about with tape recorders
tucked into their shirt pockets, cameras hidden in brown paper bags.
A TV crew filmed from the tinted windows of a van.

Some observers saw a white Animal Control truck with no license
plates trying to unload some healthy looking dogs to one of Rutledge's
regular dealers who appeared interested.

A few feet away, 350-pound Dick Garner, from Osceola, Iowa,
unsnapped his muddied overalls to relieve the pressure of the flesh,
and began dealing. Auction watchers estimated that Garner (USDA
license #42AN) sold at least four-hundred to five-hundred dogs a
month. One of Mary Warner's Minnesota contacts identified him as
one of the chief suppliers to the University of Minnesota and the
Mayo Clinic.

Garner's kennels, which housed at any given time up to two-hundred dogs, had been cited in USDA inspection reports over the years
for poor housekeeping and sanitation, and improper records that
"show more dogs than are actually in the kennel."

Al Willard (USDA license #43B013), from Codit, Missouri,
used to park near Garner. Willard had once been observed at the
Poplar Bluff auction wrangling with a pet owner who saw his "missing" dog in the dealer's truck. Although Willard lost his license in

mid-1989 for failing to pay USDA license fees, he continued selling to research until that fall. He told USDA that he did not understand he was supposed to stop dealing. His clients said they did
not know they were supposed to stop buying from the unlicensed
dealer.

On this June morning, Randall Huffstutler, a hometown boy from
Vienna, Missouri, set up shop near Garner. Randy was 250 pounds,
with curly, shoulder-length red hair and a wild auburn beard. He
looked like a Hell's Angel. Young Huffstutler took over his father,
Woody's, business—Ozark Research Supplier (USDA license
#43B047)—and kept it well-stocked for his longtime Missouri clients,
including Washington University, which he had supplied for ten
years, and the University of Missouri at Columbia.

In 1988, a local CBS network affiliate, KMO-TV, filmed Huffstutler
at the Rutledge auction buying from an Illinois Animal Control officer.
One transaction was recorded:

"Them's bait only," Huffstutler complained to the seller. He spat
a wad of tobacco into the mud by the buncher's boot. "I need 'em
forty-five to sixty pounds."

"Thems research dogs," Animal Control reassured him. "Some's
lighter, some's heavier. I got this short weight range like these two
little dogs, but there's a lot of them a little heavier or a little lighter.
Them's enough to sell."

Two years later, as USDA officials were denying that research
dealers bought at auction, Randy was still doing a brisk business, at
Rutledge, grabbing dogs by their head, jaw, or leg and stuffing them
into cages tiered on his pickup.

USDA regulations state that no more than one dog can be trans-
ported per cage in a dealer transaction, but most cages stacked atop
pickups at the auction housed at least two dogs. The interiors of these
cages were dense with feces and rotten food, and stank of urine. Such
federal violations and the lack of proper identification of the source of
these dogs would have justified severe fines and license suspension
or revocation, even confiscation of animals.

Yet the USDA—whose absence was conspicuous (their visits were
usually announced and they wore suits) had determined that all was
well at Irvin Johnston's auction. "They'll tell you what I tell you," said
Johnston. "If you take the auction out of Rutledge, it would be like
taking GM out of Detroit."

Johnston vehemently denied that dogs were mistreated at his auction or killed before closing. Still, enough complaints were called into
USDA for an inspector to pursue those charges. Dr. Edward Slauter,
Chief Veterinary Medical Officer for the South Central Sector, was
dispatched to look into the problem. Johnston was cleared of those
charges and on January 29, 1991, USDA sent him a usual reminder
notice to forward $10 to renew his license, and to make sure he had
a veterinarian on the premises during the auction. Johnston declined
to have an on-site veterinarian and, in 1991, voluntarily relinquished
his license. Nonetheless, dealing has since been observed at Rutledge
in the guise of auctioning "collars": the dogs just happen to go along.
In the ever adaptable world of dealing research dogs, this clever ruse
circumvents USDA's regulations; there is no law against auctioning
dog collars. A healthy Labrador, Hound, or any other prime research
dog was a "bonus" tossed into the deal, keeping the supply and
demand system functioning without a hitch; USDA denied this ruse
and Rutledge continued as a key marketplace, one protected by
USDA as well as by local authorities.

Theft of a pet was made a felony in Missouri in 1988, but this was
apparently news to the State Attorney General's office, which in 1990
was still not aware of that law. Shortly after the felony law passed,
House Bill No. 568 was introduced in an effort to make only "purpose
bred" animals legal for sale to research. The bill's supporters claimed
it would eliminate "random-source" dealers and so reduce, if not
eliminate, pet theft in Missouri.

House Bill 568's chief opponents were medical research lobbyists.
Representative Patrick Dougherty, who introduced the bill, recalled
the research delegates assuring legislators that they "always check
where (they) get animals from." But it was clear, Dougherty observed,
that they wanted "a cheap source of animals."

Representative Dougherty did not reintroduce the bill when it died
under the heavy artillery of medical research lobbyists. He said his
decision was not the result of any pressure from medical research
groups. Dougherty also claimed he did not receive any financial support from those special-interest groups. Truth is, however, several
medical and pharmaceutical political action committees (PACS) filled
Dougherty's campaign coffers from 1988 to 1990, among them: the
Missouri Medical PAC, Missouri Hospital Association, and the Pharmacy Political Action Committees of Missouri, all of whose constit-

uents are served by steady, inexpensive access to random-source research animals.

State Representative Vernon Scoville, now a Kansas City judge, chaired the Committee on Civil and Criminal Justice which killed the bill. Scoville's campaigns also counted on contributions from the same and other medical and pharmaceutical PACs.

Local officials also do their share to keep industry's animal-supply needs met. While Missouri citizens can register complaints against pet theft, dealers and auctions, in the northeast region a collusion between law enforcers and dealers maintains the status quo.

The official word on the Rutledge dog auction and flea market, a favorite spot for political backscratching by campaigners including, Johnston said, Missouri's Governor Ashcroft, was "hands–off."

Bill Alberty, Knox County prosecuting attorney, said in a 1990 interview that he had never been to the Rutledge auction. He saw no reason to investigate activities of the dog dealers there. All he knew was that "we've had some things arise out of there, bad checks, this and that," he told this author.

Had he looked into allegations that these were stolen pets?

"Well, now, there may have been some investigation about stolen pets, I—things don't really get to my office unless there's a case the sheriff feels like we can make. I don't know if they've checked on stolen things, but I don't think they've ever received any complaints about cruelty to animals."

Knox County Sheriff Don Bishop described the Rutledge auction site as "actually more like a flea market." He had made arrests there for possession of stolen property, he said.

Did that stolen property include dogs?

"No."

Did he have any problem with the dog-dealing transactions at Rutledge?

No, he did not.

Had he ever investigated the source of the dogs for sale?

"No."

Had he ever investigated the ownership of any of those dogs?

"No."

Why not?

"If a person wants to sell a dog, I guess he's got a right to sell a dog," Sheriff Bishop said.

What about the dogs that were dropped off on the highway, the ones that were not sold?

"Well, it's probably, um, let me see how I should answer that. I have had reports, you know, after the dog sales, about some dogs being around, yeah. I never located any of them." Sheriff Bishop said he had talked to Bill Alberty about "any cases going on in reference to dog sales."

Then he had pursued cases concerning dog sales?

"No, not dog sales."

At least one of the well-funded Missouri humane societies whose job it is to protect animals had as little interest in the auction at Rutledge. As of June 1990, Don Anthony, the longtime director of the Missouri Humane Society, had never filed any complaints with local law enforcers about dog-dealing activities at Rutledge. As of that date, Anthony had never even been to the largest dog auction in his state. The last he heard, "They just kind of closed it down. There is nothing going on at Rutledge." Anthony made this comment when dealing at Rutledge was in fact in full swing.

Missouri Humane may have had a reason to turn a blind eye to Rutledge. The animal protection society had, at one point, owned $1.5 million in stock in companies conducting animal testing. It also had had a highly profitable relationship with NASCO, the Wisconsin supplier of animals for dissection.

As for Don Anthony's pointed dismissal of Rutledge, the director of Missouri Humane had also served as Chairman of the Board of the American Humane Association, a national animal protection group which receives funds from the largest supplier of lab animal feed: St. Louis-based Ralston Purina. In its summer 1992 newsletter devoted to the topic of pet theft, American Humane relied extensively on perspective from the National Association for Biomedical Research, the lobbyists group for research institutions and dog dealers.

Arkansas television decided to visit a Missouri auction, and in May 1991 a crew from KARK-TV, Channel 4, went undercover to Poplar Bluff, the site of a bustling trade in dogs and cats of "unknown" origin. There, some of the most prestigious private and government research facilities in the South and East buy their animals.

Poplar Bluff was an experience KARK-TV reporter Mel Hanks said he would never forget.

Hanks told viewers of his series that Poplar Bluff "is a place where animals are treated like pieces of meat, thrown into the back of trucks, grabbed by the neck. It is a place where big money changes hands. Many of these animals end up in research laboratories across the country. And the place where it starts is your backyard."

While photographer Phil Pennington shot from the tinted windows of an unmarked van on a foggy Friday morning, Hanks saw "house pets handled like slabs of meat . . . puppies locked up in cages ready to be sold. The rough handling extended to terrified cats that were prodded from cages to cardboard boxes."

Wayne Tyler, a deputy sheriff from Arkansas who was interviewed for that series, said that pet protection measures do not usually dissuade dog dealers. For instance, tattooing a dog on its ear will not discourage laboratory buyers or dealers. "They just cut off that ear," Tyler observed.

KARK news reported that emotions ran high on Hanks' series. The station received hundreds of phone calls from outraged Arkansians after each of the segments. Many viewers asked why the researchers providing the market for "the scum of the earth" were not culpable. "I think they (the researchers) should be prosecuted," said one viewer. Another: "Researchers know beyond a shadow of a doubt they are buying people's pets. If you cut off the demand, then there's no need for a supply." Said another caller: "I think these people who take these dogs should be used for research. After we kick 'em."

Yet in 1991 the official word on auctions from the office of Dr. James Glosser, Deputy Administrator of USDA's Animal and Plant Health Inspection Service which is responsible for monitoring dog dealers, was:

> If someone were to be selling stolen pets (the) flea market would
> be a prime outlet for those animals. While it may be true that
> licensed Class B dealers obtain animals from this source, there
> may be other potential buyers for dogs at these markets (including)
> individuals desiring a quality pet.

SEVEN

Paper Tigers

"This is an old boy network, a popular thing. Everyone's afraid to rock the boat, it's going to hurt the good ol' boys. There are payoffs involved, politicians on the band-wagon, big money."
—*Missouri Sheriff commenting on dog dealing*

July—August 1990
America's Heartland

As dogs, many of unknown origin, were being traded at auctions that summer of 1990, Cole County Sheriff Dan Heymire was noticing a "run on Huskies" in his jurisdiction, about 150 miles from Rutledge, Missouri. Large numbers of the breed were missing from neighbor-hood yards and streets. Some weeks there was a run on Shepherds. "People go outside and their dogs are gone from their kennels," said Heymire. "These dogs don't mean a thing to the people who are taking them."

With no initiative or even interest on the part of USDA whose job it is to enforce the Animal Welfare Act, finding a stolen pet is virtually impossible. As Heymire observed, papers are forged, license plates, even dog cabs full of animals, are switched at state borders to avoid residents tracing their dogs to local research institutions. "This is a multi-level business," the sheriff said. "They have a person actually stealing the dogs, getting them from humane societies' shelters and from "Free to a Good Home" ads and then wholesaling them to the dealers. There is a lot of money to be made and it's hard to prove the person at the top has any knowledge of what is going on at the bottom."

Law enforcers across the country agreed with Heymire's assess-ment. In states where auctions were held, the numbers of dogs re-ported missing was predictably high prior to auction weekends. Former East Texas Constable Jerry Owens said that large numbers of animals vanished just days before the Canton Auction held on Mondays. He had had reports of over 1000 dogs missing from as far as Dallas during one pre-auction weekend.

In Kenton, northwestern Ohio, where "acres and acres of dogs"— as many as 4000—are sold to research on Labor Day weekend, a USDA-licensed dealer bragged, "Everyone knows guys'll go in before the sale and clear out an area and get the dogs sold before anybody knows they're gone. Nobody can't do nothin' about it." The dealer said neighborhoods were also cleaned out near Kentucky auctions in Maysville and Pikesville.

In fact, law enforcers interviewed by this author likened pet theft to the drug trade in profitability and pervasiveness. Fulton County, Arkansas, Deputy Sheriff Wayne Tyler observed, "Pet theft is so widespread, so much like the drug trade, that local residents are trying to fight back by banding together at community meetings." In Conway, Arkansas, where USDA had its regional office, eleven dogs were reported taken from one rural road alone within days. Residents in Polk County have seen pickups with empty dog cages and food scraps traveling back country roads. In Polaski County, nearly three-hundred dogs were reported missing to police within two months. Communities in LeFlore, Polk, and McCurtain, Oklahoma, have demanded local law take action, but to no avail.

Meanwhile, the USDA found auctions innocuous and downplayed community outrage. The official word from the federal agency was that "law enforcement agencies (contacted) stated they had no problems with stolen dogs or they had no record of reported dog thefts." Resident's complaints were thus dismissed.

Conclusions about auctions and law enforcer reports were reached by a USDA Task Force "established to investigate (residents') complaints that dogs were being stolen in large numbers in Missouri and Arkansas and were being sold to licensed 'Class B' dealers." Under the banner of the Regulatory Enforcement and Animal Care (REAC) division of Animal and Plant Health Inspection Services, four inspection teams were dispatched to look into allegations that dealers were buying illegally from unlicensed sources, falsifying records, failing to correct repeated deficiencies, and trafficking in stolen pets. The teams investigated the records of nine licensed dealers in Arkansas, Missouri, Indiana, Illinois, and Tennessee; conducted searches of the dealers' 148 suppliers; monitored auctions in Erie, Kansas; Joplin,

Poplar Bluff, Rutledge, and Willow Springs, Missouri. They also interviewed local law enforcers including state police and county sheriffs, and dog-pound officials.

April 1990 was targeted as the date to commence investigations by the "Midwest Stolen Dog Task Force." By then, officials reasoned, dealers would have had the opportunity to come into compliance with regulations passed in October 1989 governing dog dealing activities. According to Title 9, Code of Federal Regulations, Class B dog dealers may obtain live random-source dogs and cats from other dealers who are licensed by the USDA; from individuals who breed and raise the animals on their own property; and from pounds allowed by state law to sell to research. Bunchers selling fewer than twenty-five animals a year were exempt from licensing—but only providing they bred the animals on their property.

As for record keeping, the centerpiece of those regulations, dealers are required to maintain accurate records showing the source of each dog or cat bought for sale to research. "No person shall buy, sell, exhibit, use for research, transport, or offer for transportation any stolen animals."

Proper record keeping is the only way to document the source of any dog or cat sold to research.

Because of the seriousness of violating any of these regulations, USDA can impose severe penalties, including fines up to $1000 per count, license suspension or revocation, even a year imprisonment. If a dealer failed to correct deficiencies found during a USDA inspection of his kennels, his animals could be confiscated and his license suspended for up to twenty-one days. Failure to comply with record keeping requirements, sanitation, care and feeding, holding periods (a minimum of five days) are federal, criminal violations.

But regulations are paper tigers if they are not enforced. Thus, at auctions where illegal deals were given USDA's seal of approval, "paper tigers" wandered harmless and unheeded amid dog dealers, bunchers, and research buyers. At the close of auction weekend, these paper tigers were packed up by USDA ringmasters and carted off to the next show. In the spring and summer of 1990, the headlining act was the Stolen Dog Task Force, playing at dealer kennels and auctions throughout the Midwest.

Task Force conclusions, reached by USDA Administrator Dr. James Glosser and his protege Dr. Joan Arnoldi, Deputy Administrator of Regulatory Enforcement and Animal Care, had virtually no relevance to field investigators' reports. Over two hundred pages of field inspector notes (unreleased to the public) revealed rampant violations by dealers, including falsification of records on such a scale that even a prudent law enforcer would be compelled to act. Yet USDA's official conclusion was: "No substantive evidence was found to substantiate claims that licensed Class B dealers were dealing in stolen dogs or selling stolen dogs to research."

A random look at USDA investigators' notes concerning three of the biggest dealers in Missouri showed Glosser and Arnoldi's disregard for the evidence. For instance, the records of Randy Huffstutler, a frequenter of auctions, were examined by Task Force inspectors who found twenty four suppliers listed by the dealer during a two month period. The inspectors sought out those sellers to determine whether they were licensed to sell dogs to the research market. But after extensive searches of phone directories, post office and oftentimes police records, inspectors were "unable to locate" fourteen of Huffstutler's suppliers. As Senior Task Force investigator Harry Pearce explained to this author, "If after all our efforts we couldn't find these people, that could only mean they didn't exist." There was no "Clarence Jones" who sold Huffstutler eight dogs; no "Bill Jones" who sold him eleven dogs; no "Henry McDonald" who sold an unknown number of dogs, as did the elusive "Kermit McSpadden," "Stanley Sellers," "Robert Warner . . ."

Frank Needham, who supposedly sold Huffstutler nineteen dogs, denied selling dogs, then explained to USDA inspectors: "people brought dogs to (his) premises." Needham said that this was not the first time his name had been wrongly used on records identifying him as selling dogs to dealers.

That left six traceable suppliers who sold dogs to Huffstutler. But only one was willing to sign an affidavit indicating the source of those dogs. Nonetheless, USDA officials offered these bunchers federal license applications.

"We were told not to punish bunchers or the dealers who bought illegally, just to offer USDA licenses," inspector Pearce recalled.

USDA also observed that sixty-two dogs that were listed on Huffstutler's record were not on the premises. Randy said those dogs had been "sold" or "died." The explanation sufficed. As inspector Richard A. Gunderson wrote in his report, "The Huffstutler Family was very cooperative during this investigation," and "Randy Huffstutler may not have had prior knowledge of the revised regulations." Huffstutler was given an Official Notification and Warning of Violation of Federal Regulations and his case was marked "closed."

Huffstutler was not alone in claiming ignorance of the 1989 regulations. Arthur "Skip" Wilson, then head of the Enforcement division of USDA's Regulatory Enforcement and Animal Care, told this author that he did not know Task Force inspectors were supposed to evaluate based on the guidelines of the 1989 regulations. "No, were those the regulations?" he obliquely asked in a 1992 interview.

Despite phonied records, buying from undocumented sources, and chronic Animal Welfare violations involving inadequate housing, space, sanitation, and veterinary care, Huffstutler's Ozark Research Suppliers continues its brisk business buying dogs and housing them in what look like front-load dryers.

One of the largest dealers in the country, Ray Eldridge, was also investigated by the Task Force, as was his long time supplier, Wilbert Gruenefeld, who flaunted his knowledge of USDA codes and his clean record. Yet the business practices of both dealers were questionable.

Eldridge Lab Animals is located on Ridgefield Road, in Barnhart, Missouri, an obscure rutted road hidden in a maze of similar dirt roads off Routes 21 and M in Jefferson County. An occasional pickup truck rumbling along the gravel, a summer wind, and the chatter of birds break the dense silence of these backwoods. All that is visible of this multimillion-dollar dealer facility is an eight-foot cyclone fence, several dog trailers, a greenhouse, and black and white goats feeding in a pasture. The tips of dull green rooftops and air vents protruding above ground hint at what lies beneath this pastoral setting.

Eldridge Lab Animals Inc. (USDA Class B license #43LAH) is comprised of a cluster of football-field-size kennels stocked with as many as eight-hundred dogs each. The kennels are built underground.

Eldridge's prestige clients included the University of Missouri at Columbia, and institutions across the country, including the University of Chicago, Kansas City Veterans Hospital, and Cleveland Veterans Hospital.

While the USDA gave Eldridge a glowing report, it overlooked documented evidence that the dealer's suppliers have bought at auction, "adopted" from the "Free to a Good Home" ads, and were willing to buy from other unauthorized sources.

In 1986, a *Columbia Tribune* series disclosed that Fulton County dog dealer Ralf Bezler "adopted" dogs from the "Free to a Good Home" ads and sold them to Eldridge. One of Bezler's victims was Columbia, Missouri, resident Robin Kramer, whose seven-year-old Australian Shepherd, Merlin, was adopted by the "kind farmer." When Robin inquired about her dog, Bezler told her Merlin ran away, and he offered to help her look for him.

Eldridge's records showed that Merlin had already been bought from Bezler; the dog died in a university experiment.

Eldridge's long-term subcontractor, Wilbert Gruenefeld, of Jonesburg, was eager to talk shop. "I've got nothin' to hide," Gruenefeld said in a thick backwoods drawl. He'd had his USDA license for twenty-seven years. Number 43WG.

Did he know where the dogs he buys come from?

"You bet I gotta know where they come from. I buy 'em from people . . . different ones that raise 'em." He writes down the sellers' names, addresses, and license numbers. It's a business that "don't make me rich."

Had he ever bought from Animal Control people not allowed to sell?

"No way. Them, I don't want nothin' to do with 'em. I don't want nothin' that's illegal. I won't even take one if you give it to me. That's trouble, trouble."

In 1989, Bill Alexander, a Humane officer (now deceased), posed as a Louisiana Animal Control officer named "Jim," and contacted Gruenefeld about selling some dogs. "Jim" was not allowed to sell the dogs from his pound to research. Jim made that illegal plan clear to Gruenefeld during several phone conversations.

JIM:　　　　 I don't want my name on all of them. I can't come out like that. I'm not supposed to sell them.

GRUENEFELD: Oh, I see. (laughs) Have you got a good friend? (laughs) I tell you what. I ain't gonna lie to ya. I'm too old to start. And, uh, if you got some good friend that will let you do that,

it'll work. But otherwise, I'm not gonna take
the chance and I don't want you takin' the
chance.

JIM: I don't want to, either.

GRUENEFELD: All right. You get someone that you know
real well like a good friend of yours and use
his driver's license, tag number, that's fine.
. . . That's the way it's done.

They arranged to meet off Highway 61 near Kentucky Fried Chicken.
"I'll be sittin' over in a red Ford with a box on it," Gruenefeld told
Jim. "The box'll be brown-lookin'. You can't help but tell it's a dog
wagon. Then we'll go to a place where we can unload it."

A year later, when presented with this evidence of wrongdoing,
USDA's Dr. James Glosser, who conceived the Task Force, told this
author that "quotes are not facts," and do not constitute "evidence"
of wrong doing. No action was taken by federal agents.

When the USDA Stolen Dog Task Force investigators visited
Gruenefeld they "could not locate" several of his suppliers named on
his records. None of his suppliers who did exist were licensed to sell
dogs for research. Of the twenty-five to thirty dogs he sold Gruenefeld, one sixteen-year-old buncher said he "got the dogs from a friend."
Gruenefeld claimed he bought sixty-four dogs from a Clinton Jones
of Crocker, but Jones told a USDA senior investigator that he "did
not deal in dogs."

Gruenefeld's bunchers were nonetheless offered licenses and, despite his falsifying records in violation of federal law, Gruenefeld's
case was marked "closed."

In defense of his cronies the dealer said, "Some of these ol' boys
been in the business are about as honest as you can git. Oh, there's
a few, but 99 percent ain't half as bad as some of these people let on.
They (USDA) wished the rest of 'em would keep records like I did."

When asked by the author about the source of his dogs, Ray Eldridge slammed down the phone.

When the USDA Task Force ventured into northwest Arkansas,
the backwoods hill county that attracts dope and gun runners, and
kingpin dog dealers, its findings were as revealing.

EIGHT

Good Ol' Boys

This property is protected by shotgun four days a week.
You guess what four.
—*Sign posted outside Holco, Inc.*
(USDA License #71AA)
West Fork, Arkansas

April—June 1990
Northwest Arkansas

"Theft of pets is so common we say, 'This must be Collie day or Shepherd day,' " explained Olivia Horn, director of the Fayetteville Animal Shelter which conducts cruelty and theft investigations. The petite brunette who has endured a decade in a fast-paced, high-tension vocation, said, rather wearily, "The most common order is for male dogs, forty pounds or more, short hair, but laboratories also ask for a certain breed. It's a supply and demand business."

By law, a dealer must hold a dog or cat five days to give residents time to look for their missing pet, but few dealers will allow the public into their kennels. Owners able to gain access are often surprised at how quiet dealer kennels may be, Horn said. "Two hundred or more dogs, and you can hear a pin drop. That's because dealers cut their vocal cords with a scissor, jackknife, whatever. Researchers prefer dogs that won't whine or cry."

Unlike residents searching for their "missing" pets, USDA Task Force investigators gained access to a heavily secured Holco Kennels. There they perused records kept by Holco's owner, Don Davis, who bought a range of breeds, most in the 40 to 60 pound range, although there were those weighing as much as 100 pounds and as little as 18 pounds. Preferred breeds: all types of Hounds (Redbones, Blue Ticks, Black and Tans, Walkers, Bassets, Bloodhound, Elkhounds, Foxhounds, Wolfhounds), German Shepherds, Golden Retrievers, Beagles, Dobermans, and black Labradors. Holco also purchased mixed breeds of every sort, as well as Irish and English Setters, Pointers, Collies, St. Bernards, Australian Shepherds, Terriers, Dachshunds, Huskies, Dalmations, Weimaraners, Catahulas, Rottweilers, Pit Bulls, Spaniels, Afghans, Chows, and Greyhounds.

Holco's records showed that 21 percent of the dogs at his kennels and 38 to 50 percent of the cats died there, before reaching their research destination, from causes ranging from distemper and mange to heartworm.

The Task Force field investigators who reviewed Holco's records were "unable to locate" twenty-two of his thirty-six suppliers. One supplier who, supposedly had only recently sold dogs to Holco, had been dead for four years. Bunchers from Kansas, Oklahoma, Missouri, and Arkansas denied selling dogs to Holco. Others were buying at auction where the source of the dogs had proven to be untraceable.

These dogs and cats were sold to Holco's numerous clients, including: University of Arkansas Medical Sciences, University of Chicago, Loyola University, Theracon Inc. of Topeka, Louisiana State University, University of Missouri, University of Oklahoma and Oklahoma Medical Research Facility, University of Texas, and Shriners Burn Hospital in Galveston, to name only a few.

One of Holco's suppliers was Jerry Yaeger, a Washington County deputy sheriff and Lincoln city council member. Deputy Yaeger's sideline was discovered by reporter Debra Robinson in a *Northwest Arkansas Times* series that prompted a state police investigation a year before the USDA Task Force was conceived.

The deputy was, it seemed, "adopting" dogs and cats from the local shelters or "Free" ads on the premise he would find them good homes. Instead, he would bring the animals to Holco, which paid him $25 per dog and $10 to $18 per cat, depending on their weight.

Yaeger's now ex-wife, Carol, did not approve of the family business. Jerry's father also had, she said, been bunching dogs for fifteen years. "Jerry would drive around and pick up dogs, medium to large size, the kind the experimenters like to work with, and load them up in his dog cab. At times he'd see a Lost/Reward ad and then he'd return that dog if the reward was higher than what Holco would pay. He went through hundreds of dogs. He made about $1000 a month."

Don Davis admitted Yaeger brought him dogs, "but he was doing it himself, not as a member of the police."

Sheriff Bud Dennis denied that his deputy used his police car to pick up dogs. Meanwhile, Yaeger explained that he "trades dogs. They go everywhere, just around my home area. Just what peoples (sic) want, you know."

Despite eyewitnesses and affidavits indicating that the Deputy was obtaining shelter animals under false pretenses and "adopting" pets through the Free ads, Yaeger was cleared of theft charges and USDA never investigated further.

Holco Kennels was one of two kennels whose licenses were suspended for one year by the USDA Task Force for phoney record keeping, among other serious violations. But the suspension of owner Davis came months after he had died, in August 1990. Federal law gave USDA authority to confiscate animals in dealer kennels which were in massive noncompliance or whose licenses were suspended or revoked, but USDA took no such action at Holco. Instead, Davis' heirs reportedly delivered his remaining stock of dogs to puppy millers.

"You're talkin' about "good ol' boys," from the bunchers on up to the big shot politicians covering for each other," said Deputy Sheriff Wayne Tyler in neighboring Fulton County where the USDA Task Force checked on the records of dog dealer Dairel Caruthers, a Justice of the Peace. The Caruthers were a prominent Fulton family; Caruthers' son was land commissioner in nearby Izard County.

Caruthers, a backwoods farmer in his late fifties, sported a weathered face and dirty overalls. His dog kennels resembled most of the pens where hunters kept their hounds: wire enclosures littered with dog feces and rotted food.

Deputy Tyler explained, "A lot of these guys get licenses without any USDA inspector comin' out. All they have to do is get a vet to say 'Yeah, everythin's fine.' Even if a USDA inspector says they failed to meet requirements, all they get is a warnin' to comply. They're allowed to continue operatin' until they come into compliance, however long that takes. Meanwhile, these dealers are makin' big money."

For instance, a local county clerk was supplementing his income selling dogs at Missouri's Poplar Bluff auction. The county official was part of a consortium comprised of a USDA-licensed dealer and a realtor whose dealing kennels were on property owned by a prominent banker and judge. One auction weekend, the clerk sold two hundred dogs to a Kentucky dealer who supplied a well-known pet shop chain on the East Coast.

When the USDA Stolen Dog Task Force visited Dairel Caruthers at his Salem kennels, they observed eighteen "unidentified" (untagged) dogs on his property—a federal violation. Caruthers claimed twelve were either his own or had been "recently dropped off."

Like Holco, Caruthers also listed suppliers whom the USDA inspectors were "unable to locate," among them J. O. Turner of Sharp County. A year later, Turner was convicted by Sharp County Court of pet theft, the first conviction of a buncher in Arkansas. The action was a result of citizen outrage and the perseverance of Mary Warner's Arkansas connection, Linda Elliot.

Deputy Tyler credited Linda with changing his perspective about things he had always taken for granted.

Tyler, a strapping dark haired man with a thick mustache and gentlemanly Southern way, grew up on a farm in Batesville, and went to school with many of the bunchers and dealers he was now trying to bust. "These good ol' boys here are countin' on your thinkin', 'Hell, that's old so and so goin' huntin' with his buddies or off to some flea market trading his dogs.' Well, maybe those dogs are his, maybe they're not. You need someone to point it out to you." Linda Elliot did just that.

Years before the USDA began its Task Force mission, Linda had been monitoring dealer and buncher encampments in Fulton, Sharp, and Izard counties. "Everybody's dealin' dogs in Arkansas," she said. "We've got judges, sheriffs, police, county clerks, boarding kennels, veterinarians, and puppy millers all swappin' stolen dogs with dealers. Dealers are switchin' dog cabs at the Mississippi and Missouri borders so local dogs can't be found by their owners. These guys are travelin' back roads where they can go for miles without anyone seein' them and they're fillin' dog cabs with people's pets. You have no idea how big it is. The money is just too good."

The small-boned blond with beauty-queen looks staked out backwoods hide-outs where dope and guns were stashed along with stolen dogs bound for sale to research institutions in Arkansas, Tennessee, Missouri, Texas, Oklahoma, Iowa, Kansas, and Louisiana. Many were quality purebred Labradors, German Shepherds, Cocker Spaniels, Golden Retrievers. Arkansas kingpins pay their bunchers $8 to $25 for a dog, then resell that same animal to research for $350 to $500. For cats, dealers pay $5; $100 at resale.

For instance, CC Baird, a Church of Christ minister and dog dealer, won the highly lucrative state contract to supply dogs for the University of Arkansas and its Medical Center. There was speculation that Baird was also servicing Holco's clients, since the death of Holco's owner, Don Davis. The politically well connected CC had to have some good market. It looked like he was shipping out about three hundred dogs from his kennel every ten days; the population of Williford, his town of residence, was only about seven hundred.

In November 1991, a year after USDA had given Dairel Caruthers, one of CC Baird's suppliers, its seal of approval, Linda ventured into the encampment of yet another Baird source, the Elkin family. As she recalled, neighbors of the Elkins told her a strange odor was coming from the property; smelled, they said, like burnt, rancid meat. A sickening smell.

On the afternoon of November 23, Linda went to the Elkins' trailer outpost off Highway 9, about half an hour's ride from the Izard County seat. Some of the roughest Ozark hill people hid out in these woods, quite at home with hate groups stockpiling explosives, satanic cults, dope and gun runners, and bunchers servicing USDA licensed dog dealers. She had been met by Victor, the hitchhiker who Phyllis Elkin and her son Bobby picked up years before. They had promised to take care of the disabled Vietnam vet in exchange for his government welfare checks.

Linda suspected the Elkins were running the popular "white trash" con—bunching dogs.

"Now don't you get me nervous, don't you get me nervous," Victor warned as he tottered, bearded and filthy, at the end of the dirt road.

Linda wasn't about to get him nervous, not with his sawed off shotgun pointing in her direction and his hands trembling.

That day, the Elkins' old paramedic van had been parked near the trailer, tires deep in trash. It was usually crammed with barking dogs. Days before, the Elkins hauled a dozen, half dead animals across the Missouri line in that van. A local lawman had pulled them over, but he let them go. He later asked Linda, "What was I supposed to do?"

She replied, "What about enforcing inter-state transport and cruelty laws?"

That was just part of the trouble: well-intentioned lawmen didn't know the law, and those who did, did not bother to enforce it.

Eyeing a very nervous Victor, Linda had noticed blood on the ground and the charred remnants of a bonfire: wood, bones. She was about to leave when she saw Mrs. Elkin step out of a beat-up station wagon. The woman's stringy red hair was greasy and her bony body was covered with sores.

"What y'all doin' with all them dogs?" Linda asked her.

"We bring some of 'em dogs up Willifred way," Mrs. Elkin said. "That man CC, he buys dogs to sell for places that does tests on them."

Linda headed straight to the office of Izard County Sheriff Danny Haley.

"I'm goin' back there with a search and seizure warrant," she told Haley. "Probable cause: cruelty and pet theft." Neighbors had seen a black Lab and white Shepherd unloaded on the Elkins' property. The dogs matched the description of two dogs reported missing from Izard County, part of a rash of purebred Labs, Huskies, and Shepherds disappearing from yards and outdoor kennels. Their paramedic van had been sighted cruising those streets. Meanwhile, the local papers were flooded with ads for "lost" breeds in those areas.

Sheriff Danny Haley said, "Linda, what you gonna do is get me and my buddy killed for nothin'. They hauled out the dogs last night over to Sharp County." Haley was not about to cross over to Sheriff Sonny Powell's territory.

"Okay, Sheriff, then I'm goin' back there by myself. You know I always pack my Gucci bag with a Ruger .22 Automatic." Linda talked soft and sweet like she always did.

It took several months to get a search-and-seizure warrant. On January 9, Linda looked in her rearview mirror as she returned to the Elkins', and saw the six-foot sheriff a few yards behind her at the wheel of his Blazer. Danny Haley, a real gentleman, was protective of her. "You're a little gal and you better be scared of them people," he told her. He looked scared.

Following close behind Haley was Chief Deputy Bob Brown in the county's gray Chevy. The last vehicle in the caravan was driven by a photographer from the *White River Current* newspaper.

The dirt path to the Elkins' was littered with beer cans, plastic cups, rusted cans of pork and beans. The property was owned by one of the local land barons who seemed to delight in the upset caused by his renting to low-lifes.

There was apparently no one at home, just a few dogs chained outside the trailer and several plastic garbage bins filled with rags and papers. Junk. The Elkins were clearly on the move.

"I told you we ain't gonna find nothin' here." Sheriff Haley was anxious to leave.

Linda headed up the steps to the trailer, but a mound of human and animal feces blocked the door. She called out, "Hey, anybody in there?" The dogs within started barking.

"Okay, so you're right," Haley said. "Let's get around back."

There was a small window to the rear of the trailer. It was taped shut. Haley gave Linda a leg up and she pulled at the tape, jimmying the pane.

A bearded face suddenly filled the window. "You blond-headed fuck face! You ain't gonna take my dogs! I'm gonna kill you. Get outta here!"

It took a moment for the visitors to find their voices. Haley, who did not like cussin' in the presence of a woman, said in a tight voice, "Listen man, I'm the sheriff. We're comin' in."

"You ain't no goddam sheriff. Miss Blond Fuck Face here told you to say that. You don't look like no sheriff. And that don't mean nothin' to me, 'cept you better get yourself and Miss Fuck Face outta here."

That was it. Haley smashed his fist through the window and yanked out the Vet. The stench issuing from the trailer and Victor's body was unbearable. Deputy Brown doubled over and vomited.

The interior of the trailer was littered with feces, food and beer cans, and prophylactics. Linda pushed her way to the bedroom and shoved open the door. The dogs—she would count twenty-one— hurled toward her, more scared than mean. She could not tell what breed they once had been. Some resembled Poodles, Beagles, Terriers. Their bodies were ravaged by mange, their long nails curled around under the pads of their paws.

The dogs had toppled a tin canister and urine spilled onto her sneakers, soaking her feet.

Linda scooped up a pile of collars with Kentucky tags and tossed them outside where Sheriff Haley was retching his guts out. It took her the rest of the day to coax the dogs out. Most of them had to be euthanized, but their skin was so paper-thin the syringe could not be inserted.

When Linda filed for the search and seizure warrant to gain access to the Elkins' property, she picked up the document at the home of

Judge Connie Barksdale. Linda told the judge about the pet theft into research racket in northwest Arkansas, and how it was plaguing the entire country.

As Linda was leaving, Judge Barksdale confided: "I have to confess something to you, Linda. I did not want to be bothered with this warrant, and I was dreading your coming here. I was wrong. Animals can't speak and if we don't speak for them, who will?"

With her Southern drawl and feminine ways Linda Elliot was the quintessential Iron Butterfly, the kind of woman only the South can claim. Bunchers, dealers, puppy millers, and law enforcers on the take have called her "that old woman from Horseshoe Bend," "that bitch troublemaker," and long-standing epithet: "slut."

"If I was sleepin' with everyone those good ol' boys say I am, I wouldn't have time to take a pee," she laughed.

Fighting the Arkansas Dog Mafia was not what Linda Elliot had in mind growing up in conservative North Little Rock. Talk in the Elliot household centered on God and cops—Linda's brother was a police officer. Linda was destined to marry the right man and have babies. Her family was devout Church of Christ, and her mother was opposed to anything that hinted of frivolity.

"If I went out for cheerleadin' and the skirt was too short, I had to drop out. If I did a cartwheel and my panties showed, I had to quit. I had to quit the drill team when they switched from Bermuda shorts to skirts." When she was invited to enter the Miss City Beautiful and Miss North Little Rock pageants, her mother threw a fit. "There she goes again, wantin' to put on a bathing suit and show what she has!"

Linda had confided to her friends that she believed God would strike her dead if she even took a sip of liquor.

Coming to her rescue was her brother, sixteen years her senior. Dale Bruce had joined the North Little Rock Police Department where he would serve thirty-five years, retiring as its chief. He often took his kid sister on his cases.

The high drama of law enforcement opened a new world for the sheltered Southern belle. "There was something about livin' on the edge, the danger, intrigue, it was so different than the restrictions at home."

Linda escaped North Little Rock long enough to graduate from Oklahoma Christian College with a BA in business, despite the fact that women in the Elliot family were not expected to work. She returned home, married a local boy, and worked as a secretary. Her marriage soon deteriorated. Then, in 1974, Linda was almost killed in a car crash.

For a week she lay nearly comatose. "I prayed to God that if I could ever walk again, I'd walk out of my marriage. And I made a promise to myself to do something worthwhile."

Walk she did, with two toddler sons. Linda took a full-time secretarial job and also volunteered at child-abuse and elderly-care centers. A short time later, a fire destroyed her home and killed her dogs and cats. Then her church turned its back. "The preacher's wife wouldn't let the congregation help me because she said I was makin' her husband lust after me by mowin' my yard in my shorts."

Linda's disillusionment was profound. "Here I had done everythin' by the book, believin' that God and husband and the system will take care of you. The husband part didn't work out, the Church abandoned me, and there I was tryin' to help these abused kids and the courts kept sending them back to their parents." The crises had taken their toll.

"I guess I got my love of animals from my dad," Linda said, explaining why she attended a meeting of Arkansians for Animals.

In 1986, accompanied by an attorney member of a local animal welfare group, Linda visited the Poplar Bluff auction. She told people what she saw there and began listening in a different way to what they were telling her about their "lost pets." In 1989, a call from an elderly woman vacationing in Horseshoe Bend changed her life. "Joan McKay* told me that her Kerry Blue Terrier suddenly disappeared from her fenced yard on trash day. That dog was everything to her, the only thing she had left since her husband died."

Linda went directly to CC Baird, who was then operating one of the area's largest puppy mills. She told Baird's wife she was interested in buying a Kerry Blue Terrier. Linda recalled, "Baird's wife suddenly got real nervous and she chased me off the property."

As Linda was leaving she glimpsed a Kerry Blue Terrier in a pen, and she headed straight for the Sharp County Sheriff's office to file a complaint. "By the look on their faces when I mentioned CC's name, I knew they weren't gonna do a damn thing," she recalled.

Linda called Dr. David Sabala, USDA director of the South-Central

*Name changed to protect Subject.

region. She had had some prior contact with the agency on cruelty cases. Calling USDA was a last resort. They had consistently failed to pursue cruelty charges she had initiated in other cases, and officials once refused to appear voluntarily in one of her cases.

Dr. Sabala told Linda that CC Baird was a Class A dealer. "Then he said CC was a Class B dealer. Then he says CC really wanted to be a Class A dealer but he was applyin' to become a Class B dealer."

Later that evening a trashman/buncher showed up at McKay's door with her dog and asked for the reward. McKay told him to clear out, quick.

The case started Linda Elliot on the trail of pet racketeers in the tri-county area, a mission which led her to Mary Warner and the national criminal network. Linda explained, "This is about crime, not animal rights. It's about local law not doin' its job and the government licensin' people to steal our property."

In May 1991, Linda faced another turning point when a new resident of Sharp County called her. Lynn Patterson* was frantic: her Golden Retriever, which never wandered, was missing. "Y'all call Deputy Tyler and have him go with you onto some of those guys' properties," Linda told Lynn.

But Lynn swung into her car and headed on her own to CC Baird. Speeding up the rutted dirt road leading to Baird's house and kennels, she nearly collided with a pickup loaded with dogs.

She jumped out of her car and ran screaming toward the vehicle. "Stop you bastard! You got my dog!"

She'd know her Golden anywhere.

The pickup's driver, J. O. Turner, was not about to lose money. But this crazy woman was holding up his delivery. "You say this is your dog, lady? Then you give me $15 to git him back."

Lynn ran to her car, grabbed her purse and luckily found some money. She handed the bills to Turner.

"I'm not going to let him get away with this," Lynn told Linda and Deputy Tyler. She pressed charges, and the nightmare began. Someone left a hanging noose dangling from one of the beams in her barn. Late-night callers threatened her life. But when the case went before Fulton County Judge Jim Short, the timing worked in Lynn's favor. Within a two-week period, there had been a "run on Golden

*Name changed to protect Subject.

Retrievers" in Fulton and Izard counties, and Judge Short's own Golden Retriever had disappeared. The judge fined Turner $250. The Task Force had already marked Turner's case "closed" in 1990, with no punitive action after Turner's conviction on pet theft.

• • •

In the fall of 1990, the three page summary of the USDA's Stolen Dog Task Force was released. In addition to specifically exonerating auctions, dismissing local law enforcers' claims, and whitewashing the extent to which dealers' records were falsified, the USDA brass concluded that: "Nearly all cases showed that dealers followed the federal regulations requiring them to keep accurate records on the origin and disposition of animals used for research."

Just as USDA Administrators ignored the actual findings of its field investigators, salient evidence of theft seemed to make as little impression on the White House.

KARK-TV reporter Mel Hanks sent President Bush, in late 1991, tapes of his series on auctions and dealers. In a November 26, 1991, letter to Hanks and his photographer, Phil Pennington, President Bush responded to the station's tapes. Hanns Ruttner, Associate Director for Health and Social Services Policy, White House Office of Policy Development, on behalf of the President, thanked the newsmen for their "letter to the President and the tapes of reports regarding stolen pets being sold to licensed research laboratories."

The letter continued:

> After viewing your tape, we discussed the issue with officials of the Department of Agriculture (USDA). In 1990 USDA appointed a task force to investigate allegations that stolen pets were being sold to researchers in the Midwest through USDA licensed animal dealers. The task force examined the records of a number of suspect dealers and monitored activities at markets where dogs were being sold and traded. No evidence was found to support the allegations. USDA will continue to investigate all such claims.

Ten days later, the White House sent Hanks a letter pointing out that "the President and Mrs. Bush are animal lovers." A photo of the President with his dogs, Ranger and Millie, at Camp David was enclosed, along with the President's "best wishes."

NINE

The Open Market

"At the present time even if they have stolen dogs, as long
as they have proper records there's nothing we can do."
—*Dr. Richard Crawford, director, Animal Care,*
USDA (quoted in LA Times, March 7, 1988)

February 1988
San Fernando, California

The Ruggiero case had taken a circuitous route, but from the moment
he heard about the scam, LA Deputy City Attorney Norm Wegener
hoped that, logistics notwithstanding, this was going to be his baby.

In late February 1988, Animal Reg's Gary Olsen had approached
the Van Nuys City Attorney to file misdemeanor charges under several counts of 487G of the Penal Code: theft of an animal for sale to
medical research or other commercial purposes, which would carry a
$1000 fine; and conspiracy to obtain property by false pretenses and
promises, Penal Code Section 182.1, which carried a maximum penalty of one year in jail and a $10,000 fine.

Because of Wegener's reputation and obvious interest, Van Nuys
Attorney John Rocke called him and the file was transferred to the
San Fernando City Attorney's office. In the spring of 1988 when the
Ruggiero case arrived in the City Attorney's office, it seemed that
they would proceed with misdemeanor charges. No one could have
then foreseen that the defense would later ask that charges be upped
to felonies. The reason for this risky move was to get the case away
from the City Attorney's Norm Wegener. Wegener was notorious for
aggressively prosecuting and winning animal abuse cases.

When the file landed on his desk, Wegener's first impression was
that Barbara Ruggiero's scheme was as coldblooded and heartless a
crime as any he could imagine. The fact that these were living animals,
not television sets or any other type of valuable property obtained

under false pretenses, made the conspiracy all the more ruthless. Wegener sensed that this case was going to open a can of worms. He could only speculate as to the forces that would be brought to bear, three years later, when the case would come before the courts.

Chicago born, LA raised, Wegener grew up in a work-ethic environment. Wegener's parents were divorced; his mother ran a mortgage servicing company in LA and his father owned a cigar store in Indiana. His brother, Phillip, three years his senior, was a realtor in LA. Wegener had always been driven and felt impassioned about issues; he decided on law as a vehicle for social change.

At the University of Southern California Law School Wegener met a young woman from New York with strawberry-blond hair and cool blue eyes that matched a cool, analytical nature. Their relationship underwent various incarnations, and he and Susan Bilus remained friends.

In 1985, Wegener left a lucrative litigation practice in a large private firm in Los Angeles to join the City Attorney's office. Corporate law and Norm Wegener did not mesh, even if his salary had been four times what he made as a deputy city attorney. Not that he wanted to live his life in sackcloth. But there were some things more important to him than becoming "rich and respectable." He was just not cut out for an environment in which money leveraged justice. He explained, "I was not ethically comfortable with the fact that the 'right' side is the one who pays the bills. More than once I found that huge resources were brought to bear against an opponent who had minimal resources. Simply by way of attrition, rather than by way of practical justice, the weaker party did not prevail."

As a prosecutor for the City of Los Angeles, he had only one obligation: to serve justice. It was an excellent match.

Susan, too, had left corporate law. She had married a writer named Chasworth and joined the DA's office to represent the People as a deputy district attorney. Her office was now on the third floor of the San Fernando Courthouse; his was on the basement level. In the spring of 1988, Susan was doing the Juvenile Court rotation, one of the toughest stints. Kids killing kids, child molestations, gang wars.

Wegener heard that Susan's marriage was faltering, but he knew she would weather the storm. It was her nature to remain somewhat detached; he tended to get more emotionally involved.

A year into his practice, Wegener began taking on what were considered the toughest, most volatile cases, those that merited even fewer of the office's scarce resources. These were the animal cruelty cases where, more often than not, the victim was dead.

Wegener had always loved animals. He had two dogs, a German Shorthaired Pointer he had rescued from the streets of LA where the dog was starving, and a Shepherd-Lab mix. Prior to the Ruggiero case it would have never occurred to him that there was even a market for your average family pet, let alone a value of $500, the price medical researchers were willing to pay someone like Barbara Ruggiero. Had he been forced by circumstances to give up his own dogs, he might have been one of Ruggiero's victims.

"I thought about the dogs that were accustomed to living in a home, to the loving and nurturing of an owner, that were suddenly shut up in a laboratory cage and periodically drugged and operated on, treated like a commodity, an instrument. I could only think what a horrible, horrible trauma that would be. If I had been one of her victims, I would have been outraged to the point of vigilante violence."

The first cruelty case he tried concerned a Sylmar man who shot a stray foraging in his garbage. Wegener tried the case in front of a jury, a high risk, but he won; the culprit served ninety days. Although the sentence did not meet Wegener's expectations, it was considered a coup by his office.

By the spring of 1988, the Deputy City Attorney had tried twenty cruelty cases and won jail sentences for all the offenders.

Now he was facing a case that both morally and emotionally repelled him. Yet the psychology of Barbara Ruggiero fascinated him and he was anxious to examine her on the stand. He had learned bits of her background from a private investigator hired to gain more information about the case; the details were bizarre. Douglas Cameron, whose specialty was criminal investigation, had interviewed Ralf Jacobsen in February 1988. Ralf, it seemed, was repentant, willing to talk and anxious to get Animal Reg and the police off his back.

The clean-cut, blond young man who introduced himself to Cameron certainly did not look like the monster capable of committing such cold-blooded deceptions. It was clear to the PI that Ralf felt "he was going to get his butt in the sling." His ex-lover and her current fiancé were trying to put all the blame on him, the odd man out. He said the media was unjustly focusing the spotlight on him and his life was being ruined; his military status with the National Guard was in jeopardy. He had already lost his job at a law firm and he was afraid of being thrown out of the San Fernando Valley Law School.

Ralf told Cameron that he was receiving anonymous threatening phone calls, people saying he was "a slimy bastard, a butcher." Someone had poured sugar in his gas tank. He was being followed by persons unknown; they said they were going to get him.

It had begun as a college romance. Ralf first met Barbara Ruggiero in September 1980, when he was attending Pierce College. They began dating in December, and in March 1981, they moved to Fresno where Barbara gave horseback riding lessons and raised rabbits for meat and pelts. They broke up in late 1981, then reunited briefly in 1985.

Bridgette Jacobsen was not happy about her son's obsession with Barbara. Years later she explained, "At the beginning I liked Barbara. I thought she was the sweetest girl. But the devil comes in nice faces. She was a very manipulative person. I could see she was a user. She practically told me to go fuck myself."

During 1985, Ralf and Barbara began their first joint business venture. Ralf told Cameron, "We would answer an ad someone put in the paper for Dobermans needing good homes. We would go to the person's house and act like we were a couple wanting the dog for our pet. We'd get the dog and Barbara would give it obedience and guard dog training, and we'd sell it."

Then, in mid 1987, Barbara met Rick.

Ralf, the spurned lover, hovered nearby, willing to accept any scrap of affection she'd throw his way.

In October of that year Barbara asked Ralf, who was then working for a law firm, if he were interested in making some extra money. She told him she needed as many large dogs as he could find and advised him to go through the *Recycler* ads, as they had done with the Dobermans. But this time she told him not to use his real name and to make no promises or guarantees to the pet owner. As far as the pet owners knew, he was simply adopting the dogs for companionship.

After each delivery of ten pets he would receive a check: $10 for each cat, $20 for each dog. Barbara would sometimes go out on her own, or make arrangements to meet him with the van and load up the newly acquired pets. She said it "wouldn't look good" if there were more than one or two dogs or cats in the car when they were going to a pet owner's house.

"I got cats from the ads," Ralf said. "But I took them from the streets too, the ones that were running loose. Barbara knew about that. Sometimes she told me to take in dogs running loose in the neighborhood. That happened a few times. Once, I remember, there was a black Lab she grabbed right in front of Budget. She took off his collar and put him in the kennel."

Ralf said he had some qualms about what he was doing. "I felt bad sometimes taking animals from their homes. It was like taking a child away from his family. I told Rick how I felt and he seemed to agree. Barbara never said anything."

In November, while going through the *Recycler* in Barbara's office, Ralf noticed a list of hospitals tacked to the wall. He asked her about the names. "They were the hospitals she was selling the animals to. They were going for medical research. I was surprised, but she said she had a government license to do it. Rick said he spoke with a government attorney who said they weren't doing anything improper as long as they kept accurate records."

Rick, the business manager of the dealer kennel they called Biosphere, which was located at the site of Budget Boarding, sent out letters to 125 institutions licensed by USDA to perform animal research. The letters stated, in part:

> Our main goal is to provide good quality, clean and parasite dipped animals to laboratories and research facilities. We supply domestic dogs and cats presently, but plan to expand our selection to include other laboratory animals in the near future.
>
> We provide and guarantee animals of sound quality. . . . If you have special facility requirements, as to the nature or qualification of the animals, we will be glad to work with you to try to meet them. All orders will be filled as promptly as possible, usually within two weeks. However it would be wise to order as far in advance as possible.

Rick had spoken with UCLA and other institutions to determine competitive prices. They decided on:

Cats: $100
Adult dogs: 0–20 pounds: $150.00
21–40 pounds: $250.00
41–74 pounds: $350.00

75–99 pounds: $400.00
100+ pounds: $500.00

One of Biosphere's first visitors was Dr. John Young, Director of Laboratory Animals for Cedars Sinai and consultant on Lab Animals for the Veterans Hospital in Sepulveda. Young advised Rick against posting any signs indicating the kennel was selling to research and warned Rick about "animal activists."

Rick would later testify that he also spoke to a staff veterinarian at the USDA's Sacramento office who told him that obtaining animals for research from "Free" ads was legal.

Ralf told Cameron that he cut back for a while when he learned where the animals were going.

"Why didn't you stop, walk away?" Cameron asked.

"I wanted to be around Barbara. I guess I was blinded by love."

On one occasion he accompanied Barbara and Rick on a delivery to the Sepulveda Veterans Hospital. They drove a load of ten cats around back where a doctor met them. The doctor rejected two cats because of eye infections and signed some paperwork for the remaining eight. Another time Ralf and Barbara went to the San Dimas pound and adopted nine or ten dogs. "We told them we were going to find homes for the dogs," Ralf explained.

Wegener thought it must have looked innocent enough, and he felt for the unsuspecting victims. The Deputy City Attorney did not consider himself an animal rights activist, but he did believe that too much research went on. It seemed to him that its only purpose was to justify researchers' salaries.

"The medical community is very powerful, very entrenched. It doesn't have any incentives to look for alternatives, and certainly does not have an incentive to complicate the process of obtaining animals by asking where these animals come from."

The stakes in the Ruggiero case were high. Wegener knew he would be up against some formidable opponents intent on diverting the real issue away from theft and deception and into animal research and animal rights. He anticipated that the defense team would leverage public sentiment against animal rights activists. The well-oiled public relations machinery of the research industry would undoubtedly contend that the dealers were performing a valuable service. Wegener's predictions were not far from the mark.

His first task was to remand Ralf Jacobsen, who was then in the reserves at Fort Bragg, a military base in North Carolina. Extradition posed a problem. The City Attorney's office would not allot money to apprehend the suspect in another state. Wegener contacted the base commander and explained the situation. The officer said, "If we were at war, Mr. Wegener, I'd have told you to go to hell. But since we're not, we'll hold Jacobsen. But you'll have to send someone to get him."

In mid-March, Animal Reg's Gary Olsen and Bob Penia were dispatched to North Carolina and returned with a terrified Ralf Jacobsen.

That month, over sixty victims of Ruggiero's theft ring had been identified. There would be at least eighty more victims who would be named, and potentially hundreds of others whose pets had been acquired before Biosphere was even licensed. These animals had allegedly been sold to a USDA-licensed dealer in Oregon who then sold them to Cedars Sinai and other West Coast institutions.

On September 8, 1988, six months after the case landed on Wegener's desk, Barbara Ann Ruggiero, Frederick John Spero, and Ralf Jacobsen were charged by the City Attorney's office with eleven counts of petty theft, eleven counts of theft of an animal for medical research purposes, nine counts of conspiracy to steal an animal for medical research, and conspiracy to obtain property by false pretenses and promises. Barbara was also charged with one additional count of theft in connection with receiving $35 from Chuck Ransdell as an adoption placement fee—instead, she had sold his dog, Ammo, to Cedars Sinai.

USDA's reaction was predictable. On October 7, 1988, less than a month after the City Attorney filed charges, the agency sent Biosphere a letter reminding the dog dealers to renew their Class B license. USDA sent Biosphere yet another reminder letter on January 29, 1989, when the theft and conspiracy charges were upped to felonies and Barbara and Rick were out on bail.

Meanwhile, USDA showed quite a different face to the LA public, which was panicked about its pets being stolen for sale to local research institutions. In March 1988, Dr. Richard Crawford, senior staff veterinarian at USDA, promised that USDA would harshly prosecute criminal dealers and "strive to return stolen pets to their owners." The agency, in fact, did neither.

Crawford apparently had developed amnesia two years later in 1991, when he responded to inquiries about the USDA Task Force

Report: "Nobody is really sure where stolen dogs are going to. There is really no evidence yet that points to their going into research. In fact, if you try to track them down, there is really no evidence that many of them have been stolen."

•   •   •

Just as the Ruggiero case hit the California press, Dr. James Glosser, Administrator of the USDA's Animal and Plant Health Inspection Services, was creating a new subdivision of his agency—Regulatory Enforcement and Animal Care. REAC would presumably concentrate on enforcing the Animal Welfare Act with its main objective to crack down on pet theft. Glosser named Dr. Joan Arnoldi, a former State Veterinarian in Wisconsin as Administrator of REAC, and issued statements that pet theft was a priority issue.

Back in Virginia, Mary Warner saw this effort for what it was: a public relations maneuver. She anticipated that the new division, REAC, would have as little impact on pet theft as any other government entity had. She had been faced with the scope and potency of government corruption fifteen years before Barbara Ruggiero learned about the lucrative business of dealing dogs and cats to research. Mary had formed Action 81 with that harsh reality in mind.

When Fate brought the Godfather of the Pennsylvania Dog Mafia virtually to Mary's doorstep in 1976, she learned just how entrenched this insidious theft network was.

TEN

Kingpins and Capistranos

"People were ripping dogs off the street and the main person they sold to was Esposito."
—*Officer Nathan Jenkins, Virginia State Police*

February 1976
Martinsburg, West Virginia
Interstates 64 and 81 run parallel, the more northern 64 crossing Kentucky and West Virginia, while 81 cuts through Tennessee and the western part of Virginia. At Harrisburg, Pennsylvania, they converge and hurtle toward Quakerstown, the longtime home of Sam Esposito's Quaker Farm Kennels.

Interstates 64 and 81 "belonged" to Sam Esposito, boss of the Pennsylvania Dog Mafia. It was along these routes that his "capistranos" picked up hundreds of thousands of dogs and cats from auctions, pounds, streets, and bunchers who met Sam's trucks at collection sites off the highways. Those transports hauled the animals to Quaker Farm's football-field-size pens and outbuildings nestled in the stark Amish landscape. There former pets, some starving, sick, dying, awaited shipment to Sam's clients. According to an internal delivery list, those clients included: the National Institutes of Health, University of Pennsylvania, Yale, Johns Hopkins, Princeton, Rockefeller University, Sloan Kettering Cancer Institute, Pittman Moore Paints, U.S. Surgical Corporation, Mt. Sinai School of Medicine, North Shore Hospital, and Ortho Pharmaceutics.

On February 24, 1976, a West Virginia state trooper stopped to assist a pickup truck that had broken down on Interstate 81, twenty-five miles from the home of Mary Warner. The trooper was shocked when he observed the contents of that vehicle: nearly a hundred dogs, some sick, others dead. One dog would later be identified as a "missing" local pet.

The driver, who was from New Jersey, was charged by police with cruelty, transport violations, and failure to identify the dogs he said he had acquired en route. He paid the $63 fine and phoned his boss, who then instructed him to call Chambersburg, Pennsylvania.

"Sorry I can't help you," was the response from Russell Hutton. "My trucks are broke down."

The driver made another call. "My boss is sending a couple of trucks," he told the authorities. "These animals got to get to a laboratory in Pennsylvania."

"You always carry so many dogs?" a police officer asked.

"Hell, Sam's told me to fit in two hundred in the pickup sometimes. But I tell you the dogs don't always get to where they're going alive."

The trucks belonged to Sam Esposito. The call for help was made to Sam's buddy, dog dealer Russell Hutton. Esposito and Hutton respected each other's territory and thus avoided territorial disputes. But in a fix they would help each other out.

This incident, thereafter referred to as the "Martinsburg incident," marked the beginning of Action 81. Mary, then sixty-four years old, rallied a few Animal Control officers, local humane societies, and hundreds of victims. "The numbers of people who had actually witnessed a theft were astounding," Mary recalled. "There were so many missing dogs and cats in the same or adjacent neighborhoods. The stories I heard, grown men crying on the phone, saying their Labs, Shepherds, hunting dogs that never wandered suddenly disappeared. I remember one woman was gardening and her Doberman was outside with her. She turned around and saw a truck swoop past and her dog was in it!"

As Action 81 gained local notoriety, Mary received a macabre message. Someone deposited a dead Shepherd at the gate of her farm. The dog's inner ear had been scraped, but enough was left of the tattoo markings to convey the sender's message—that no effort at dissuading bunchers or dealers would work.

A year later, in March 1977, Mary received another, more encouraging message. "I'm with the Dog Law Bureau in Pennsylvania," the man on the phone, who identified himself as Roger Musselman, said.

Mary vaguely knew that the Dog Law Bureau in Pennsylvania was a division of that state's Department of Agriculture and was under the federal USDA. Musselman said he had a tape she should hear. Mary thought he sounded scared.

It was a particularly cold morning when Musselman and two fellow agents with the Dog Law Bureau delivered the tape to the Georgetown home of Faye Brisk. Faye, one of Mary's colleagues, was a former press secretary to Lyndon Johnson and a valuable conduit to Washington power brokers.

Musselman, Tom Hammer, and Tim Brennan explained that an investigation they began in September 1976 had been shut down. The officers said their jobs were on the line. They had been harassed, threatened, betrayed. "We don't know what it's all about, but it seems there's a lot of people who are very nervous about what we're finding out," Brennan told Faye and Mary. "We feel we're tugging at loose strings. We hoped you'd be able to clue us in."

The investigation concerned what appeared to be a dog larceny ring in Pennsylvania, a pet-theft-into-research racket involving Dog Law Bureau personnel.

It began, the agents said, with a tip from a reliable informant implicating two Dog Law wardens who were stockpiling dogs. The agents audited the wardens' records and found large numbers of dogs unaccounted for. Then, suddenly, Dog Law Supervisor Larry Harteis ordered the agents off the case.

Musselman said, "We, the three of us, met with the Assistant District Attorney to see if we could get some action." They did, and the Cambria County police filed charges against one of the wardens for "theft by deception and falsifying records."

Steve Lewis, the Bureau Director, told the agents to continue, and that the investigation was now going to be under his direction. Mary was immediately suspicious. She had recently received a letter from Lewis concerning Esposito's picking up dogs illegally in Virginia; Lewis had defended the dealer.

By mid October, the agents' surveillance of the wardens had led to kennels run by Russell Hutton and William Clark, both USDA-licensed dog dealers. "We heard rumors about people gathering dogs and selling them to research kennels," Brennan said. "It looked like we had something. The dealers' records were atrocious. Half the records had no previous owner listed or just listed Clark or Hutton as the owner. At both kennels we found no records of where the dogs went. Steve Lewis told us that illegal dog traffic would be easy to hide."

Within weeks the investigation had gone into other jurisdictions. In another county, police filed charges of forgery and receiving stolen goods against the dog warden who supplied Hutton. Supervisor Harteis again tried to stop the investigation, which had by now landed in the DA's office.

On December 1, the agents met with officials from the Dog Law Bureau and USDA. The tape Musselman slipped into the recorder at Faye Brisk's house was of that meeting.

Musselman's voice on tape: *"Esposito's shipping out eight hundred to nine hundred dogs a week to big clients. Research is asking for specific breeds—Irish Setters, Shepherds."*

Tim Brennan continued, *"The United Kennel Club in Michigan says there's ring of stolen Coonhounds. Dealers are making bogus USDA tags."*

"Look, boys, the purpose of the Dog Bureau is to license dogs. That's it. Why don't you stick to what you're supposed to do?" The speaker was Frank De Garcia, a Dog Law Bureau official who worked under Steve Lewis.

The USDA veterinarian present at that meeting agreed that the agents should back off.

"It's bigger than just this state," Musselman retorted. *"Minnesota dogs are ending up in Pennsylvania. Veterinarians are rubber-stamping health certificates for interstate transport. Good dogs are disappearing."*

"It's true," Steve Lewis said. *"Pennsylvania dealers are going to Virginia for dogs."*

Mary felt her stomach tighten at Lewis's comment.

The tape played on. Musselman told the USDA vet, *"Your agency is not cooperating with us. You're not doing inspections. We need someone to get into the labs and backtrack to regions where the dogs are coming from."*

The tape finally came to an end. Mary and Faye sat for a moment in stunned silence. Then Mary asked, "What happened after that meeting with USDA?"

"We were getting to that," Brennan smiled, but his voice was tight. "Five days later we got new orders. The investigation stopped. On December 8, we met with Lewis, De Garcia, Harteis, and our regional director. We were told: (a) We had spent too many hours and too much overtime on this investigation; (b) We had opened a 'can of worms' which was reaching into other regions and would sap our manpower; (c) It was not the Bureau's responsibility to investigate dog larceny; and (d) Dog dealer kennels were taken from our jurisdiction. Lewis washed his hands of it."

"Then who was responsible for inspecting dog dealers?" Faye asked.

"A Bureau vet who was selling veterinary supplies to the dealers," Brennan replied.

The agents had given their report to the Secretary of Agriculture. "He never read it," Musselman told the women.

Mary suggested a meeting be arranged with Dr. Frank Mulhern, then USDA Administrator responsible for dog dealers. She immediately phoned Christine Stevens, founder of the Animal Welfare Institute, whose fortitude had help pass the Animal Welfare Act.

The "Mansion," as the Stevenses' home was known, was a Georgian manor sited on an entire city block in the prestigious Georgetown section of Washington D.C. The neighborhood had the particular charm of cobblestone streets banked by the immaculate townhouses of the Washington power elite. Within, the mansion sported the well-worn look of gentrified wealth: antique Persian rugs, select Chippendale furniture, Oriental art, Impressionist watercolors, overstuffed velvet sofas, high-backed upholstered chairs, floor-to-ceiling bookshelves, with books stacked also on chairs and tables.

The basement of the Mansion was reserved for the small staff of the Animal Welfare Institute. Over the decades, Christine held fast to her belief that USDA was simply ignorant of dealers' corruption, and she maintained a friendly and hopeful dialogue with USDA. She and Mary often locked horns over their dissenting views of USDA's culpability.

Christine had access to government power brokers and, within days, Dr. Frank Mulhern was sitting in her formal dining room. The three women, the Dog Law agents, and the USDA Administrator listened to the tape. Mary prayed Mulhern would order his own investigation.

"Um, um," Mulhern said, shaking his head as the tape concluded.

USDA never took any action against the suspect dealers, and the

Dog Law Bureau agents eventually left to pursue other law enforcement positions.

When asked in 1992 about his meeting with the Dog Law agents, Mulhern said he could not recall the session. He had no recollection of the tape.

•   •   •

Looking back on the Dog Law Bureau's early days, Kathy. Schweitzer, one of Mary Warner's Pennsylvania contacts, explained, "The Bureau meant shit. Everybody knew Sam Esposito had them in his pocket. The Bureau was only formed to cash in on the Amish puppy mill trade by requiring those factories be licensed. Dog Law got a perk by getting extra money from the dealers."

Kathy, thirty-four, a straight-talking Humane officer, had been working the animal abuse beat for fifteen years. She became Mary's connection in the late 70s, Esposito's heyday. For decades, Sam Esposito had dominated the East Coast dealing scene with his political coteries and payoff procurement network that extended to the heartland of America.

"In Pennsylvania, Sam got an easy slide on Dick Thornburg's Administration," Kathy said of then governor Thornburg, who went on to become U.S. Attorney General under Ronald Reagan.

Thornburg's appointees to the Pennsylvania Department of Agriculture reflected the Administration's mentality. Between 1979 and 1984, Penrose Hallowell served as Pennsylvania Secretary of Agriculture; the Bureau of Dog Law was under his jurisdiction. Hallowell was heir to the family business, Hallowell Labs and Penrose Labs, suppliers of laboratory animals. "Penny was this multimillionaire and he was caught shoplifting a Barbra Streisand tape," Kathy said.

During Hallowell's stint as Secretary of Agriculture he named Don Moul head of the Dog Bureau. "We couldn't get Moul to do a damn thing about the dog dealers or the auction," Kathy recalled. The system was already deeply entrenched in the state's economy.

By 1987, the USDA-licensed lab animal business in Pennsylvania was the richest, fastest-growing in the country. Hazelton Research Products, Inc. grossed $3,540,000 in sales to research institutions; White Eagle Laboratories, $1,138,208; Esposito's Quaker Farm Kennels, $1,097,729; the Buckshire Corporation, $1,044,176; Biomedical Research Associates, run by Mike Kredovski, Esposito's competitor, $908,665; Haycock Kennels, Inc., $730,535. The remaining four of the state's top ten dealers each grossed about a quarter of a million that same year.

By the early 80s, the tri-state territories of Pennsylvania, Virginia, and New Jersey were divided up among the big-time dog dealers: Rudolph and Helena Vrana (USDA license #22B21) in New Jersey; Noel Leach (USDA license #52BA) in Virginia; and a saturation of multimillion-dollar facilities in Pennsylvania. Many of these dealers were members of Esposito's network, supplying him with dogs and cats and delivering for him in a crunch. As such they were immune to prosecution through Sam's connections with USDA, the Dog Law Bureau, and various state legislatures.

The Vranas supplied New York hospitals including Mount Sinai and Albert Einstein College of Medicine, and U.S. Surgical Corporation. In 1983 and 1985 they were convicted by New Jersey courts of violations of the Animal Welfare Act, ranging from filthy conditions to transport violations, including ninety dogs covered with feces and crammed into a pickup. Pet theft eventually did them in. In 1988, the couple was thrown out of New Jersey by the state courts on charges of stealing pets.

Nonetheless, the Vranas' loyal clients defended the dealers. Dr. L. J. Serrano, director of the Animal Institute at the Albert Einstein College of Medicine in New York, said the Vranas "were just being harassed because they were emigrants. They did such an excellent job of taking care of their animals."

Yeshiva University, of which Einstein is a branch, received grants from the National Institutes of Health in excess of $68 million in 1989. Einstein currently obtains dogs from another New Jersey Class B dealer who "ends up going to auctions, as they all do," conceded Serrano.

In Catawba County, Virginia, Noel Leach, one of USDA's first licensees, was raking up years of violations, ranging from deplorable sanitation and housing to animal transport violations ("dog on truck in a cage so small it could not stand. The animal was filthy with its own excrement") and allegations of pet theft corroborated by pet owners' sworn testimony that they had found their "missing" dogs at Leach Kennels. The USDA's response to twenty-six years of documenting violations at Leach's kennels and citizen complaints was to issue one "cease and desist" order, a warning which kept the dealer in business.

Leach's clients included Eastern Virginia Medical School, George Washington University and Carolina Biological Supply, which sold dogs and cats to schools for dissection classes and where, the public would later learn, cats, many allegedly stolen pets, were embalmed while still alive. Leach also supplied A. H. Robins, the pharmaceutical company that was liable for millions of dollars in lawsuits arising from their faulty IUD, the Dalcon shield.

Vrana and Leach bought at auctions where Esposito had his contacts, including those in Kentucky where he set up a satellite kennel. Gathered around Esposito's truck at the bustling Paintsville auction, bunchers lauded the dealer, who was "doing the public a favor getting garbage dogs off the road."

In Pennsylvania, the Roots and Gilbertsville auctions were popular buying spots for dealers. There, pet marketers traded information on who was buying and what was selling. Officer Kathy Schweitzer also frequented those auctions, on the lookout for violations that would stick.

"Sam and Lester were always together, shootin' the shit at Roots," she recalled.

Lester Young was a Pennsylvania Dog Law Bureau inspector and Sam Esposito's hunting buddy. Young's vigilance in cracking down on unlicensed dogs was a boon to Esposito, especially in the summer when the Bureau prosecuted "negligent" dog owners who did not license their pets. Young told the local press that he believed pet owners would rather get $3 selling their dog to Esposito than pay $5 for a license.

"Sam and his son Joe hung up their sign and they paid three bucks for a healthy adult dog, two bucks for a cat," Kathy said. "They'd charge laboratories $250 and up for those same dogs, $100 for cats. Sam had a good thing going."

No one asked too many questions at the Pennsylvania auctions about the origin and destination of dogs and cats purchased there— except Kathy Schweitzer. In August 1982, she was forcibly removed from the Gilbertsville auction for being too inquisitive. "A couple of henchmen from the auction and the Dog Bureau escorted me out. I came back in. I don't take 'no' easy. They smacked me around a couple of times in the face. So I went to the Dog Law hearings with a black and blue face."

Kathy was one of eight children growing up in a working class home in Reading, Pennsylvania. Her father was a truck driver, her mother a housewife. When she was eighteen, Kathy noticed a newspaper ad for a caretaker at the local animal rescue league. "I thought, this is it. This is what I always wanted, to be around animals." She was quickly promoted to Humane Officer and finished college with a degree in criminology, enabling her to become an expert witness on civil and criminal cruelty cases.

What she saw as an investigator toughened her for the dog auction beat. Common sights were "dog shit knee high," and emaciated horses dying of thirst, their hooves so overgrown that when the animals stood up "they looked like a dog when it takes a dump, that's how contorted they were."

Kathy soon heard about the auctions, that they were "shithouses," and she went to have a look. Through that circuit she met Sam Esposito. They chatted frequently, often ribbing each other to see how far they could go. "I once saw Sam at a sale of health food store equipment. I asked him, 'Sam, what the hell is a dog dealer doing at a health food auction? You buying health food for your dogs?' He told me he was buying the refrigerator trucks to transport the dogs. He said he was going to keep the health food logo on, to keep a low profile on the road."

Esposito once offered her a wad of several hundred dollar bills. "The idea being I would look the other way. I told him, 'no thanks, Sam. I could use the money, but I'd rather wake up in the morning and be able to look at my face in the mirror.' I think he respected that and he never tried it again."

Sam often bragged to Kathy about his steady client, the multibillion-dollar National Institutes of Health. NIH had publicly claimed that it bought only purpose-bred dogs—that is, dogs raised specifically in breeding colonies for research.

"Sam said it was a bunch of bull about NIH not using random-source dogs. He'd been supplying them for years. He said he was concentrating on the D.C. area so he'd stay out of Mike's territory.* They had an understanding about that. Sam bragged that if it hadn't

* Mike Kredovski, another USDA-licensed dealer.

Before

An East Coast supplier of laboratory dogs, prior to USDA's being given responsibility for enforcing the Animal Welfare Act and licensing dog dealers, 1966. (STAN WAYMAN. LIFE MAGAZINE © 1966 TIME WARNER INC.)

After

A Midwest supplier of laboratory dogs in 1986, twenty years after USDA began enforcing the Animal Welfare Act and licensing dog dealers. (BOB BAKER)

The Laboratory Supply Business Today

Opposite, top: Dog dealers at a recent Rutledge, Missouri auction. (CRAIG SANDS, KANSAS CITY STAR) *Opposite, bottom:* A USDA-licensed dealer set-up at a Ripley, Mississippi auction. (BOB BAKER) *This page, top:* A typical USDA-approved dog cab at Rutledge, Missouri, 1990. (CRAIG SANDS, KANSAS CITY STAR) *This page, bottom:* A USDA-licensed dog dealer loading up his truck with over one hundred dogs for the 600-mile trip from Missouri to Minnesota in 100° weather. (PROTECT OUR PETS)

how we parent our

FREE TO GOOD HOME
LOVEABLE. BLK
LAB/SHEPHARD MIX. FML
2 YRS. PHONE 0000

en's
ity
s t
H(
ia
(

to hear of the death of
eland of Houston who

After
half

Background: A typical "Free to a Good Home" ad.

Top: Candy Sheker with Pooches on the left, Wiggles on the right.

Middle: Norman Flint with Fred, Cody and Wiggles. Bear died in an experiment at Cedars-Sinai.

Bottom: Collars of dogs and cats acquired through "Free to a Good Home" ads for Cedars-Sinai, Loma Linda, and Veterans Hospital, Sepulveda, California. (JUDITH REITMAN)

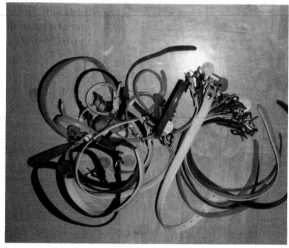

Top left: Barbara Ruggiero in court. (PETER CANATTA)

Top right: Los Angeles Deputy District Attorney Susan Chasworth. (PETER CANATTA)

Bottom: From left, Ralf Jacobsen, Barbara Ruggiero, attorney Lewis Watnick, Rick Spero, and attorney Eli Guana in court. (PETER CANATTA)

Opposite, top left: Mary Warner in the "Dog House." The map on the wall shows the heaviest pockets of pet theft around the country. (JUDITH REITMAN)

Opposite, top right: Mary Warner and her "rescues" at home. (LEON WARNER)

Opposite, bottom: Conditions at a North Carolina USDA-licensed dealer of cats. (PEOPLE FOR THE ETHICAL TREATMENT OF ANIMALS)

Above: CC Baird's USDA-licensed kennel. (LINDA ELLIOT)

Left: Linda Elliot rescuing dogs from a buncher encampment supplying USDA-licensed dealer CC Baird in Arkansas. (WHITE RIVER CURRENT)

Left: A random source dog at the University of Washington laboratory. (Progressive Animal Welfare Society)

Middle: Random source dogs in the laboratory (Physicians Committee for Responsible Medicine)

Bottom: "Rex," one of 18 dogs stolen by a Minnesota buncher for the Mayo Clinic and University of Minnesota. He was found at the Mayo Clinic, 1984. (Humane Society of U.S.)

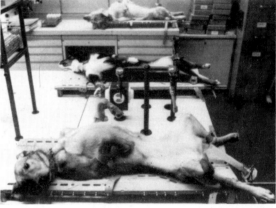

been for him supplying dogs to NIH we wouldn't have birth control pills. He was selling hundreds of dogs to them a month for cancer research."

Esposito also sold dogs and cats to Federated Medical Resources in Chester County, Pennsylvania. FMR was a coalition formed by the University of Pennsylvania in the late 60s, comprised of five medical schools planning to conduct animal experimentation. The other members were Temple University, Philadelphia Medical College, Thomas Jefferson University, and Hahneman Hospital. FMR decided to breed its own dogs and cats because of what it complained were "restrictions" imposed by the Animal Welfare Act. However, after three years in business, FMR stopped breeding and began buying cheap dogs from Esposito, housing them in substandard on-site kennels. FMR meanwhile lived rent-free, on a multimillion-dollar state bond. When the coalition was finally evicted fifteen years later, it owed the state $4 million in back rent. That amount was foisted onto taxpayers.

"No one will talk about Federated Medical and the state," said Kathy. "There was a lot more going on there than politicians want you to know about."

•   •   •

Sam Esposito held as much sway in Mary's home state as he did in Pennsylvania. His power base in Virginia became evident when the "Martinsburg" incident finally went to trial in 1978. USDA assured the public that the minimum penalty would be a thirty-day license suspension and a fine; but in court, USDA officials suddenly backed down, deciding that the evidence against Sam was really not that strong.

"Some dogs were uncomfortable," USDA wrote in defense of Esposito. "Some were hurt and a few were dead."

Esposito was found guilty of a few reduced charges and prohibited from driving his truck for fourteen days. USDA explained: "The suspension should not be so long as to seriously threaten the existence of his business."

Sam's defense counsel in that case was Raymond R. Robrecht, Delegate to the Virginia House Assembly. Another highly placed employee of Sam's was Senator William A. Truban. Truban, who was also a veterinarian, served in the Virginia Senate from 1969 to 1991, when he opted not to seek reelection. Esposito hired the senator to inspect his vehicles for interstate transport and issue health certificates.

In 1980, Truban had successfully lobbied on Esposito's behalf against a bill which would have banned out-of-state dealers like Esposito from access to Virginia pounds. A year later, the senator was prohibited from voting on the proposed bill, pushed by Mary Warner, because of his relationship with Esposito; the bill passed.

Just a few years earlier, a Virginia state police investigation of Esposito had implicated Senator Truban and other dealers as part of Sam's network. Nathan Jenkins, then chief police investigator, recalled: "People were ripping dogs off the streets and the main person they sold to was Esposito. These dealers knew each other, just like in the drug business. They were getting heavy money and going after a better breed of dog."

Jenkins presented USDA with evidence of pet theft in Virginia, including a tape that described a criminal network along the East Coast whose boss was Sam Esposito. Several years later the tape was returned to the state police. The incriminating segments had been deleted. USDA took no action.

In subsequent years, Esposito and his foot soldiers were charged with cruelty, transport, and health violations throughout the Virginias, but still USDA took no action. When, in 1983, similar types of charges were filed in Missouri, USDA again defended Sam Esposito. The Office of the Prosecuting Attorney for Phelps County, Missouri, noted: "An expert witness from the USDA indicated that there were no violations." The plaintiff, Missouri Humane, did not even bother showing up.

As for the fate of Quaker Farm Kennels, after Sam Esposito died, reports varied. Some sources claim his stock and equipment were purchased by several large dealers; other sources said they have since seen Esposito's son Joe at Midwest dog auctions.

ELEVEN

Foxes in the Henhouse

"It makes my blood boil to think taxpayers are paying this
Agency to do a job they are not doing. Why should an
old couple on Social Security pay for this farce?"
—*USDA inspector, granted anonymity*

There was a short pause and the phone connection crackled with
static. The operator intervened. "Please deposit forty-five cents for
an additional . . ." The enforcement officer was calling this author
from a pay phone. Calling from his office was too risky. "They found
out there was a 'leak,' " he said nervously. The "leak" was information
documenting USDA turning a blind eye to dealers stealing pets. "It's
such a damn mess. They don't want anything let out."

But why would the government want to protect criminals?

"Big money and powerful careers," he said. "A lot of not-so-nice
secrets that need to be kept. Politics and politics."

Dr. James Glosser and Dr. Joan Arnoldi were conspicuous players
in this political drama. Each had been a State Veterinarian working
for the government: Glosser in Montana's Department of Agriculture
and Arnoldi in Wisconsin. Each had been industry advocates, favoring
a hands-off policy when it came to enforcing the Animal Welfare Act
in their states' agriculture and research industries. In return, industry
favored them.

Animal disease was Dr. James Glosser's supposed forte. Heavy-set,
in his early sixties, Glosser had a long-standing preference for research
over veterinary practice, which he found too stressful. In the mid
1970s, he joined the Epidemiological Investigation Services, a specialty group of the Center for Disease Control in Atlanta. He remained
at CDC for eight years, studying diseases transmitted from animals
to humans. It was at CDC where Glosser made his high-level govern-

ment and industry contacts whose later support would prove invaluable.

It was also at CDC where he acquired his philosophy of "monitored self-regulation," which would be a boon to dealers and the research industry. Since CDC is a government research facility, it is not obliged to report to USDA the numbers of animals it uses, how it uses them, or their source. CDC was supposed to monitor itself, a task which repeatedly proved ineffective given the number of its animal deaths due to poor care and inadequate monitoring.

Glosser's coup came in December 1983, when he was made Assistant to the Administrator of the Animal and Plant Health Inspection Services (APHIS), a $271 million branch of the USDA then headed by Burt Hawkins. Through APHIS Glosser could expect to gain national prominence, and the sort of prestige which would ensure his future employability in the more lucrative private sector of his choice: the research industry.

"Heading up the Animal and Plant Health Inspection Services is akin to having been a senator in how employable you are afterwards," observed a government insider. "APHIS is a career move." Another government source anticipated that the research industry "would welcome with open arms a top USDA official whose policy was one of obstructionism to regulation." That source explained, "Under Glosser, research and its use of animals gained a rather malignant attention."

Glosser, it seemed, served as if his job was to protect dealers and their research clients from too close scrutiny.

One of Glosser's first moves as Assistant to the Administrator was to try to rid APHIS of its assigned task of enforcing the Animal Welfare Act, a job it historically resented. He proposed that funding for Animal Welfare Act inspections be reduced or eliminated. When that proposition failed to win congressional approval he complained that APHIS just did not have *enough* money to enforce the Act. The contradictions did not strike him.

Later, Glosser suggested that the responsibility for enforcing the Animal Welfare Act be assigned to the National Institutes of Health, the major funder of animal-research projects. That idea also did not go over well with Congress, since House subcommittee hearings concerning medical research fraud at NIH were then underway.

NIH was being investigated in 1988 for its inadequate and incestu-

ous peer review system, which allowed fraudulent research to flour-
ish. In his syndicated column Jack Anderson quoted an NIH
whistleblower who said that at NIH: "The peer review system does
not pick up fabricated work. You can get away with murder for a
pretty long time." NIH's "self-monitoring" was, in fact, *covering up*
rather than exposing fraud and abuse. It was "protecting offending
scientists and putting whistleblowers on the defensive." NIH's investigative panels were themselves staffed with scientists conducting animal research "who are hesitant to lower the ax on their peers."
Evidence of scientific fraud was also mysteriously disappearing. In
1979, a drug researcher under investigation said all his data was lost
when his "rowboat flipped."

Despite Glosser's efforts, the Animal and Plant Health Inspection
Services was stuck with the Animal Welfare Act. But Glosser made
it clear to his staff that animal welfare was not a priority, and, in
1985, when he became Associate Administrator and Acting Deputy
Administrator of Veterinary Services, he required his veterinary medical officers to spend minimal time on inspections of dealer kennels
and research institutions. And when violations were found, prosecut-
ions were rare.

Glosser philosophized that violators "should be led into compliance
through education." USDA not only allowed violators to remain in
business, it offered them tutorials on how to circumvent the government's own regulations. For example, USDA encouraged researchers
to "use stock phrases from the APHIS instruction memorandum" to
justify to their Animal Use Committees the withholding of anesthetics
from animals during painful experiments.

The APHIS Glosser joined had been faring poorly. At the end of
1983, the General Accounting Office (GAO), which evaluates government agency performance, blasted APHIS for its abysmal enforcement record. Only twenty-one of the fifty-six violations cases that
involved research facilities that year were resolved, and in all cases, *no*
punitive action was taken. There had been *no* follow-up to determine
whether those facilities corrected their deficiencies. In fact, in the
coming years many of those same institutions and thousands more
would have chronic, severe federal criminal violations endangering
the health of their animals and their employees. The Animal Welfare
Institute's review of voluminous USDA documents found conditions
like those at the University of Virginia, where dogs were housed

in barrels, their bedding soaked with feces and urine; at Harvard University, where dog bedding was covered with feces and food waste and no sanitation system was in place; at Johns Hopkins University, where monkeys were found "wet and smeared with excretia." And at Michigan State University there were dogs standing over open cesspools. At the University of Minnesota "bags of dead animals and waste paper" littered a laundry room where a washing machine was covered with bird droppings. Cat boxes were full of old fecal material. The University of Virginia had adopted a housing idea from the dog dealers: dogs were kept in barrels where their bedding was covered with feces and soaked in urine. USDA continued to license these and hundreds more facilities in violation.

Among the universities in chronic noncompliance for failing to properly identify dogs and cats during one year alone of Glosser's reign: Mayo Clinic, Washington University, the five sites of the University of Georgia, University of Hawaii, Johns Hopkins, Michigan State University, Ohio State, University of Pennsylvania School of Medicine, Oregon Health Sciences University and Medical Center, and University of Minnesota.

Meanwhile, industry's successful lobbying against the mandatory use of anesthesia resulted in graduate students in Professor Russell De Valois's psychology class at University of California at Berkeley operating on conscious kittens during painful surgery and using surgical techniques that were "unsanitary." These violations prompted a former clinical veterinarian at Berkeley to comment: "The problems of UB laboratory animal care have always been a lack of will or interest on the part of those in authority." At the University of Pennsylvania, baboons underwent head trauma surgery without adequate anesthesia. Facilities at the university's School of Medicine were in "very poor sanitary condition"; cages were too small and rusty and roaches crawled about.

In its evaluation of APHIS, the General Accounting Office concluded: "Despite the fact that a finding of major violations indicates that there may be danger to the animals, there appears to be no system to protect the animals."

A year later the U.S. Office of Technology, which also assesses federal agency performance, condemned APHIS as being "literal and cautious with regard to (enforcement at) research facilities."

It was a mind-set which Glosser found not only commendable but worth encouraging when, in April 1988 he became Administrator of APHIS. In Dr. Joan Arnoldi, whom he was to appoint head of his newly created subdivision, Regulatory Enforcement and Animal Care, in 1988, he found the ideal compatriot.

•   •   •

Wisconsin in the mid 80s was Joan Arnoldi's training ground for federal employment. There she served as State Veterinarian, before assuming her prestigious administrative reign at the U.S. Department of Agriculture. Wisconsin was also home to Class B dog dealer Ervin Stebane.

In the summer of 1986, Appleton communities were stunned by a *Post Crescent* ten-part, sixty-article series of horror stories about Stebane's Circle S Ranch reminiscent of an article written by another local paper about Circle S *twenty-six years before* that detailed the shocking conditions there.

Public anger and Congressional pressure forced Dr. Arnoldi to investigate. It was an uncustomary move since during her first two years as State Vet she had rarely, according to her successor, enforced any Animal Welfare regulations. But what the series reporter, Jim Flasch, and what Arnoldi concluded was markedly different.

In 1986, ex-employees of Circle S told reporter Jim Flasch that puppies were routinely and cruelly disposed of, and that the dealer castrated dogs without anesthesia. "You don't have to waste your money on no anesthesia," Stebane reportedly told his workers. "Just chop it off."

Those witnesses spoke of cages reeking of urine "so bad we couldn't breathe"; of feeding dogs meat that had "big wads of hair, like a fifty-pound slab . . . the dogs would fight over"; of a barn where "a whole load of other dogs were kept. Looking down a hole . . . we could see a couple of dead dogs, laying on the ground and dogs all over them. The dogs were half eaten."

It had been twenty years since USDA first licensed Stebane, and still the horror camp behind his pristine white farm house was, "one hell of a shock."

The series that rocked the quiet Wisconsin community began as a fluke. Flasch stumbled onto the story, which was to have been a short local-interest piece. Mrs. Ervin Stebane, a longtime Appleton resident, was entering Appleton Medical Center (AMC) for open-heart surgery. Flasch had heard that her husband supplied dogs to AMC for research which would ostensibly save his wife's life. "A heartwarming story," Flasch later recalled with an ironic smile.

But when the amicable reporter called the research center for some details, the institution flatly denied using dogs.

Something didn't sound right. Flasch got suspicious and did some of his own research.

He requested USDA inspection reports from the Freedom of Information Office and learned that Stebane not only supplied Appleton Medical Center but the University of Wisconsin, Fromm Labs, and Madison Area Technical College. In one USDA inspection report, a teacher from the Madison Technical College had complained about the condition and source of Stebane's dogs. The USDA inspector noted that Dr. Robert Taylor told him, "Stebane (is) stealing dogs, sells them to MATC . . . real sick, mange, parvovirus, tick infested, parasites." . . . Taylor later denied making those statements.

While flipping through stacks of reports about Circle S Ranch from 1980 to 1986, Flasch noted that USDA inspectors had reported sixty-seven life-threatening (to animals) violations, but had taken no action against Stebane. Government records also cited at least two incidents in which Stebane received dogs from "Free to Good Home" ads; still USDA took no action. The reporter spoke with residents, who told him that Stebane was using a stun gun to steal dogs from streets and yards. In one instance, a nun watched helplessly as the dealer's van made off with her dog.

During his investigation, Flasch learned about Action 81. He began a correspondence with Mary Warner about the problems he was encountering. He wrote to her that Dr. Ron McLaughlin, USDA vet in Madison, was interfering with his inquiries. "McLaughlin has both hindered and covered up any investigation into the work of Ervin Stebane. I believe he is tipping off the dealer."

Wisconsin Congressman Toby Roth demanded that the Office of Inspector General look into Circle S. "In most cases the USDA either looks the other way, or merely slaps Mr. Stebane with a simple fine and the inhumane conditions continue," Roth told Jim Flasch.

Public anger forced the office of State Veterinarian to act. On July 9, Dr. Arnoldi sent two of her investigators to Circle S Ranch. This was an uncustomarily aggressive move. As state investigator, Ron

Paul, explained, under Arnoldi "there was no enforcement at all. I was hired to correct that."

And what did the State Veterinarian's office find in its investigation? In the spring of 1992, Dr. Joan Arnoldi, by then head of the USDA division whose job it is to monitor dog dealers, was asked about her findings in that sensational case. She replied, "I don't remember."

The state had, in fact, found no problems at Circle S Ranch. Stebane's kennel was "just about the same as any other kennel in Wisconsin."

Several months later, Flasch disclosed that Arnoldi's office had sabotaged an undercover investigation of Circle S by local residents.

Through it all, Stebane's clients stood resolutely by him. The University of Wisconsin's own investigation found no problems at Circle S Ranch. The school, which received about $90 million a year in federal grants, depended on Stebane's dogs and cats for dozens of research grants, among them a $200,000 yearly grant that involved destroying the visual cortex of cats (blinding them) in an attempt to understand why brain damage impairs eyesight. "This is a standard 'ablation' study which has been going on at Wisconsin and dozens of other institutions for years," commented Dr. Brandon Reines in his evaluation of that proposal. Dr. Reines, a medical historian, author, and consultant to the Medical Research Modernization Committee, an organization of health care professionals critical of animal research on scientific grounds, said of that type of research, "That's how experimental psychologists earn a living. Relevancy to human disease is irrelevant."

Another U of W experiment, this one on a $120,000 a year grant, studied the damage caused by injecting nerve poison into dog's intestines.

"This is philosophical research that has no human applicability," Reines observed.

Nonetheless, incomes were at stake, and with the help of his loyal clientele, Stebane rode out the public storm.

Meanwhile Joan Arnoldi's maneuvering had caught the eye of Jim Glosser. They clearly shared the same philosophy about enforcement, the solidly built woman with her signature short crop was named Glosser's assistant pending the launch of Regulatory Enforcement and Animal Care. REAC's goal: "to provide leadership and professional services to achieve optimum compliance with the laws and regulations that govern the health and care of animals and plant resources and protect the public interest."

•　　•　　•

Joan Arnoldi had an immediate opportunity to show her loyalty to her new boss, Dr. Glosser; a crisis had arisen in Oklahoma in the fall of 1988 concerning a USDA employee turned whistleblower, Harry Pearce. Pearce, a long-time USDA inspector, had already gone to the federal Office of Inspector General with proof of a cover-up in the state/federal Brucellosis program. Brucellosis is a highly infectious disease prevalent in cattle which causes abortions, sterilization, and loss of milk productivity. The disease is also transmittable to humans who drank infected cow's milk. Brucellosis poses perhaps *the* major economic threat to the livestock industry. Cooperative efforts to eradicate the deadly disease were filling state and federal coffers with ten of millions of tax dollars each year.

Yet, unscrupulous livestock managers were often hastening to market Brucelloslis-infected cattle in hopes of still earning a profit, and they were sometimes aided by USDA inspectors. Pearce and another whistleblowing colleague, USDA veterinarian Ralph Jenner, had proof of such collusion in Oklahoma, where Brucelloslis was still rampant. A subsequent Office of General Counsel investigation supported their claims.

Brucellosis was a politically crucial issue for Glosser. He had made his name while Montana State Veterinarian by ostensibly eradicating the disease in that state. Disclosure of a scandal in Oklahoma would imperil Glosser's showcase program. It was clear that immediate intervention was required before this local spat became a full-blown federal war.

In the summer of 1989, Joan Arnoldi and her staffer, Dr. Arthur "Skip" Wilson, then head of the Regulatory Enforcement arm of REAC, flew to Oklahoma. There they met with State Agriculture Commissioner Jack Craig.

The officials contrived to bar Pearce from access to his office's files. "They were afraid I might blow the whistle on other things," Pearce recalled. "But if I couldn't get to the records, I couldn't do any investigations."

By late summer, Arnoldi had demoted Pearce and reassigned him to Kentucky because of "considerable tensions" in the Oklahoma office.

But Pearce fought back. He appealed to the Office of Special Counsel and got a ninety-day stay. Oklahoma Congressman David McCurdy also petitioned USDA on Pearce's behalf. The local press was meanwhile exposing the federal/state cover-up and Pearce's treatment by officials.

To avoid further embarrassment, Arnoldi made a deal with Pearce. He was removed from Brucellosis assignments, would work out of his home in Oklahoma until his imminent retirement, but he would have no access to office files on the disease.

Arnoldi's smothering of the Pearce incident gave Glosser the protection he needed. Criticism of the internal workings of APHIS had been escalating for years. Investigations conducted in the late 80s by the Office of Inspector General had found that the Animal and Plant Health Inspection Services was plagued by money mismanagement and unethical behavior. The OIG concluded that "fraud and collusion" was rampant, and that the centerpiece of corruption was the Brucellosis program, Glosser's bailiwick and biggest money-maker. "If the true conditions of Brucellosis in the herds came out," said a USDA inspector in Texas who asked to remain anonymous, "it would show USDA has not accomplished its mission."

OIG investigations disclosed that APHIS employees were covering up incidences of diseased herds, falsifying test results, and condoning substandard animal quarantine conditions that were inciting disease epidemics. In 1983, for instance, USDA approved an airport shipment of parrots carrying Newcastle disease. The fatal disease then rampaged through U.S. poultry populations, destroying millions of chickens. In 1988, when Glosser would assume the helm of APHIS, USDA inspectors approved severe health and sanitation violations at a horse quarantine facility, deficiencies that the Inspector General concluded "could have resulted in the dissemination of foreign animal disease in the U.S."

A year earlier, the Inspector General found that APHIS's Brucellosis Information System, launched with $6 million in taxpayer funds plus millions more for maintenance, was "too complicated and costly" to run and of little benefit.

The General Accounting Office disclosed other internal problems at USDA including employee embezzlement, insurance fraud, and bribery. The Inspector General also cited two USDA inspectors who were selling cocaine at their stations, and others who were charged with sexual harassment. The President's Council on Integrity and Efficiency gave USDA similarly low marks. It found that USDA agencies had not reported about $7 *billion* in income payments made in 1984 and 1985.

The public remained ignorant of these goings-on, but the upwardly mobile Glosser knew he needed a strong ally. In Dr. Arnoldi he had found someone who was willing to play ball.

Glosser's friendship with Joan Arnoldi was forged from a shared pro-industry philosophy and shared professional associations. As State Vets they both had served on the Board of the U.S. Animal Health Association (USAHA), an industry group in Richmond, Virginia, which makes recommendations to the government on the Brucellosis program. Its constituents include federal and state veterinarians, the National Cattlemen's Association, pharmaceutical companies, and kennel operators.

"USAHA is not a stomping group for virtue," conceded one of its members. "You are there to serve the interests of your industry and to make sure government knows what you want done."

This intent provided the basis of Glosser and Arnoldi's partnership.

●　　●　　●

By 1988, stolen pets and the horrifying scenes at dog dealer kennels were familiar headlines in the regional press, from Pennsylvania to Wisconsin to California. In 1984, Bud Knudsen's kennels in California were busted by a San Joaquin County sheriff and the squalor and pet theft revealed by the press. That same year, a Mower County, Minnesota, sheriff tipped off residents that their stolen dogs might be at the Mayo Clinic and the University of Minnesota. With the help of Carol Kough, the wife of the mayor of Austin and a close friend of Mary Warner, sixteen pets including Ella Erie's Sheltie, Rex, Wayne Soskow's Shepherd, Sheba, and Dave and Mary Grignon's Retriever, Isaac, were found at the Mayo Clinic. Two other stolen dogs had already died in University of Minnesota experiments which were conducted even though the university suspected the dogs were stolen. All the stolen dogs found at the Mayo Clinic were in ill health; one had to be destroyed.

In 1984, the Mayo Clinic was using approximately 2000 dogs a year for such projects as the "Teflon" experiment. The Teflon experiment involved injecting fourteen female dogs and four male monkeys with a radioactive Teflon preparation to observe the movement of that substance through their bodies. Four of the dogs had to be killed ten days into the experiment and discarded because researchers found their radioactivity level was "too intense" and would skew the results. The researchers concluded that the effects of Teflon on lab animals "raises concern of (its) use in humans."

Then, in late 1987, a worker at Sam Esposito's Quaker Farm Kennels which serviced some of the most prestigious universities and hospitals blew the whistle. He had kept a diary during that summer which documented the trafficking of thousands of dogs and cats into Sam's kennels from auctions in Tennessee, Kentucky, and Missouri, and from undocumented sources in Illinois, Iowa, and the Carolinas. The animals were arriving dead, sick, and maimed. There were at any given time at least 1500 dogs and five hundred cats living under substandard, inhumane conditions awaiting shipment to clients in Connecticut, Delaware, Maryland, Massachusetts, New Jersey, New York, Ohio, and Pennsylvania. A sample of his diary:

> *May 20:* Dogs in new arrival pens kicked in head . . . open sores on feet and pad areas.
> *May 21:* Water still not changed. Found one dead dog, appeared to have been bitten severely by other dogs—eyes gouged out.
> *May 29:* Four dogs found dead while feeding today.
> *August 6:* Dog that had chain around neck apparently killed.
> *September 5:* Pickup truck with TN plates brought approx. thirty-five small to large dogs—no food, or water, no shavings. Two dogs near dead. Dogs coated with urine and feces.

And on and on. The diary was given to People for the Ethical Treatment of Animals, the animal rights group in D.C., which held a press conference charging the dealer with over two hundred alleged violations of the Animal Welfare Act, pet theft, and state cruelty violations actionable through the Pennsylvania Attorney General's office. The disclosures made headlines in the tri-state area and were carried nationally through the wire services.

The Pennsylvania Attorney General took no action. Neither did USDA. Memorial Sloan Kettering Cancer Center and Rockefeller

University came out publicly to defend their supplier, and shipments from Quaker Farm continued uninterrupted to Sam's clients.

U.S. Surgical Corporation, the country's leading manufacturer of staplers used in surgery, was demonstrating its product on about 1000 live dogs bought each year from dealers including the Rudolf Vranas and Quaker Farm. U.S. Surgical CEO Leon Hirsch claimed the dogs he bought "arrive with their own patent charts. It is all according to federal regulations. We know everything about these dogs. You would get exactly the same care as our dogs do if you checked into the Bridgeport Hospital." Hirsch refused to discuss Esposito or the Vranas.

As part of their sales training, U.S. Surgical's layman staff perform surgical procedures on an average of seventeen live dogs in what the company calls "dog labs." One procedure involved slitting open the dog's stomach lining, inserting the staplers, and then trying to tear apart the incision by pulling at the animal's stomach lining. Other procedures these nonmedical salespeople performed involved removing the intestines, stomach, colon, spleen, and kidneys, then stapling back the incisions to close the inflicted wound. The dogs are then destroyed.

Ironically, such "dog labs" have already been discarded by many medical schools since alternatives to animals exist for those procedures. Among U.S. Surgical's critics was Dr. Alfred Cohen, chief of colo-rectal surgery in the Department of Surgery, Memorial Sloan Kettering Cancer Center, which, like U.S. Surgical, bought dogs from Sam Esposito. Cohen, who is also an associate professor at the Cornell Medical School, claimed that U.S. Surgical's use of dogs to instruct its sales staff to supposedly "train surgeons" is "unnecessary, cruel and ultimately not in the best interests of human health care."

Dr. Cohen uses the staplers, but has never attended U.S. Surgical's "dog labs" or trained on any dog in his career. "Dog labs are nothing more than clever sales gimmicks on the part of U.S. Surgical to sell their products in a very competitive environment," he said. "Dogs are not the flight simulators of the surgical world, and the argument that surgeons must first practice on dogs is fallacious. Surgeons learn by observing other surgeons and by being supervised doing the actual procedure on humans." Cohen explained that a dog's intestinal track is not analogous to a human's. "If anything, the surgeons who 'practice' on dogs must relearn the feel of human organ response. That in itself is a setback." His stance was supported by 30,000 physicians who later signed a petition denouncing U.S. Surgical's "dog labs." Leon Hirsch refused to accept the hand-delivered appeal.

In 1987, the year of the Esposito disclosures, Oregon residents learned that S&S Kennels on Highway 20 in Lebanon vied with Quaker Farm in the severity and numbers of violations, including pet theft. S&S Kennels (USDA license #92B50) was a West Coast counterpart of Ervin Stebane's Circle S Ranch in Wisconsin, where REAC's future administrator, Joan Arnoldi, first flexed her administrative muscles.

Filth and disease festered for over two decades at S&S Kennels, which were run by James Hickey, another one of USDA's first licensees. Cats and dogs stolen from yards and streets, and others taken from "Free to Good Home" ads were identified by their owners at Hickey's kennel. Several local bunchers were convicted by municipal court of stealing pets and selling them to Hickey. Illegally acquired Greyhounds were also found at the kennels.

These dogs and cats awaited shipment to drug companies and academic institutions including Cedars Sinai and UCLA. However, for twenty-five years, USDA renewed Hickey's license, defended the dealer, and thwarted residents' attempts to claim their stolen pets.

Linn County was fed up and its local court took action. County enforcers determined that Hickey was committing severe health and care violations and "concealing the source of his dogs." As would occur innumerable times when local law enforcers acted against dealers, USDA was embarrassed into action. Riding on the coattails of county court enforcement, USDA Administrator Jim Glosser announced that Hickey's license would be suspended for twenty-five years and he fined the dealer $40,000, the stiffest penalty awarded to date. But that amount would never be collected by USDA.

For years after, USDA would point to the Hickey case as a sterling example of its enforcement. What the public did not know was that USDA allowed Hickey to remain in business another year while he appealed his sentence—and while his son Joe applied for his own license. Not only would father and son continue as a team operating the kennel under substandard, criminal conditions, but Joe incurred serious violations on his own. While Glosser was lauded for Hickey Senior's suspension, his own staff was defending the son. Dr. Richard Overton, a USDA veterinarian who would later be promoted to a regional supervisory position, told the Linn County press, "Mr. (Joe) Hickey is, we've determined, acquiring animals legally. He is not intentionally doing anything wrong. I think sometimes Mr. Hickey is getting a bum rap. We have to be careful of the witch hunt syndrome."

Eventually Joe, too, was convicted by a local court and voluntarily gave up his USDA license.

•   •   •

During the Hickey furor, the "Sosha" case in the state of Washington entered the judicial system. Donald Johnson, a computer programmer in a Seattle suburb, needed to find another home for his German Shepherd mix, Sosha, and he diligently advertised for a "good home only." With the help of a local front-line organization, the Progressive Animal Welfare Society, Johnson learned that Sosha's "good home" turned out to be the University of Washington, where she died in an experiment studying the effects of pneumonia on lung disease. Sosha had been "adopted" by a buncher.

Johnson decided he "did not want Sosha to die in vain." He was suing a University of Washington dog dealer for theft. The university, which receives about $114 million in tax money each year from the National Institutes of Health for its research projects, was blaming USDA.

APHIS was in a tough spot. On the heels of public and congressional pressure, it was forced to take a stand. The challenge was to assuage public ire over pet theft without antagonizing APHIS's research constituents. This delicate political maneuver was managed by publicly admitting the existence of pet theft but exonerating the research industry of any blame. This coup was accomplished by the proposing of regulations to a law which sounded promising on paper but had, in reality, little chance of passage in light of anticipated research industry reaction and its power over APHIS.

TWELVE

Corridors of Power

"The people of this country have a right to protect their
pets and prevent them from being stolen."
—*Senator Wendell Ford*
Senate Hearings on the Pet Theft Act, April 1988

The Office of Management and Budget—the name hardly conjures
up potent images of White House power-brokering. Yet it was behind
the closed doors of the OMB, one of the least known yet most powerful
agencies of the federal government, that the fate of over 120 million
family pets was decided in the spring of 1987.

Two years earlier, Kansas Senator Robert Dole and California Congressman George Brown had proposed that the Animal Welfare Act
be amended to address the still rampant problem of pet theft. The
amendments would specify the legal sources of dogs and cats for
research and better physical and psychological conditions for animals
at research laboratories where violations thrived unchecked. The
amendments passed in 1985, tacked unobtrusively onto a Farm Bill.
The next step was for USDA, the agency that administers the Animal
Welfare Act, to propose "regulations" concerning just how this new
law would be enforced.

Without regulations, laws can languish for years. The amended
Animal Welfare Act did just that—until USDA was backed into a
corner and forced by public outrage to propose regulations.

Still, it would be many more years before those proposed regulations would pass. In fact, the law to protect the American public and
its pets was delayed, sabotaged, and finally subverted by medical
research lobbyists, dog dealers, and USDA's top brass in the seques-
tered Office of Management and Budget.

The federal government relies on regulations as its basic tool to
spell out how a law will be implemented. Regulations determine

programs and policies and can be as important as the law itself. During the process by which regulations are made—a process called "rule-making"—special-interest groups try to parlay their influence to, in effect, get the best deal for their constituents.

Unfortunately, the objectives of those special-interest groups may not reflect the public's interest. Government officials may have hidden agendas and industry lobbyists may have special access to decision makers. The result is that the public's welfare is often forfeited in backroom deal-making.

Special interests swung into action in March 1987 when USDA's proposals were published in the *Federal Register*, a government publication that announces proposed regulations.

> In the past few years there have been several instances of licensed dealers obtaining dogs and cats by fraudulent means and apparently knowingly purchasing stolen animals. The department has also noted an increase in unlicensed dealers buying dogs and cats at flea market or trade day type sales . . . from anyone. The net effect of the above types of activity is to encourage animal theft for profit.
>
> We certainly do not believe that a research facility would knowingly purchase stolen animals, but after such animals have gone through a dealer, there is no way to tell if such animals were possibly stolen.

The regulations specified where dealers could legally obtain dogs and cats for sale to research:

> Class B dealers may obtain live random-source dogs and cats only from state, county, or city owned and operated pounds and shelters.* The Class B dealer would not be able to obtain random-source dogs and cats from non-government, contract or humane pounds or shelters, or from individuals who did not breed and raise the dogs or cats on their own premises.

The regulations, which required strict record keeping on the part of both dealers and research buyers, would, USDA assured the public, "effectively block the growing activity in stolen and in fraudulently obtained dogs and cats."

*Those pounds and shelters that could legally sell to research.

Thus, twenty years after the Animal Welfare Act was passed, regulations concerning the source of dogs and cats for research were proposed by the USDA. But in a move seemingly calculated to confuse the process and make its monitoring more difficult nonindustry advocates, the proposed regulations were divided into three parts, each part with its own separate publication date in the *Register*. One White House observer commented, "The volumes of proposals and their fragmentation was a deliberate attempt to prevent supporters of the law from ensuring that effective regulations would be issued."

USDA's next step revealed just how much it wanted those regulations passed. The agency went straight to the biomedical research industry and formally requested its comments on restrictions to its trade. USDA even went to dog dealers asking them for comments. In tandem with the National Association for Biomedical Research, the industry lobbying group, USDA sent industry's predictably negative comments to the Office of Management and Budget, an historically industry-friendly agency which has the authority to approve or reject regulations for virtually all federal agencies.

Citizen groups, most notably OMB Watch in Washington D.C., have been attempting to draw public attention to this little known yet powerful entity. OBM has the staggering authority to determine the annual budget of almost every federal agency; withhold funds appropriated by Congress, thereby freezing spending; and control all regulations by telling federal agencies when to undertake rule-making, what the substance of the rules will be, and how the rules are to be implemented.

Agencies must get permission from the Office of Management and Budget before engaging in *any* activity which may someday lead to a regulation. OMB conducts its reviews in complete secrecy so that the public learns only after the fact, in an often censored and incomplete form, the outcome of the rule-making process. OMB not only can disapprove any agency request, but it also oversees government information policy, deciding what information the government should collect and who gets access to that information. OMB also directs the internal operations of agencies, including financial management, and has used its powers to privatize government activities.

The Office of Management and Budget has such sweeping powers that no federal agency can support or oppose legislation without obtaining its clearance. In fact, *any* negative feedback from OMB's reviewers could justify amending, even aborting, proposed regulations. Scores of proposed regulations on industry have thus fallen into what David Vladeck, a lawyer with the Public Citizen advocacy organization, called "the black hole of OMB." Among the rules proposed for the public good, then diluted or swallowed up into oblivion at OMB: auto industry air pollution and safety regulations, pre-treatment of hazardous chemical wastes dumped into public waters, warnings on children's aspirin, safety rules for underwater divers, steel industry air pollution standards, and cotton-dust controls for the textile industry. Hundreds more regulations that would have cost industry money but protected the public's health and the environment were similarly dismembered by OMB reviewers; so too was the fate of the USDA's proposed regulations for the Animal Welfare Act.

An indignant National Association for Biomedical Research swept into the very receptive Office of Management and Budget claiming that government—specifically USDA—was overstepping its authority. The medical research industry was going to protect its multibillion-dollar investments and put that offender back in its place.

First off, NABR objected to what it called "rigidly detailed procedures" imposed by the regulations. Those offensive procedures related to animal care (including exercise for dogs in laboratories) and acquisition (limiting the sources of dogs and cats). The proposed regulations were, industry spokesperson Frankie Trull argued, "time-consuming, costly, and unduly burdensome without any corresponding benefit to animal welfare."

Trull, the head of NABR, which was founded by the largest lab-animal breeder in the country, wrote to USDA in defense of dog dealers: "We are told by several of the affected dealers that 70 to 80 percent of the animals they supply come from sources to be declared unallowable by APHIS proposed regulations. The Class B dealers affected are likely to be put out of business." In effect, industry was admitting that up to 80 percent of dog dealers were buying from potentially illegal sources.

The American Council on Education, many of whose members use animals in experiments, worked aggressively with NABR to change the regulatory approach of USDA. One former OMB staffer was particularly helpful in this regard. Industry hired a former counsel to the Office of Management and Budget, Michael Horowitz, now a private attorney who still wielded considerable influence at OMB. Through NABR's connections, industry found a way to deflect the law's intent.

In official correspondence with USDA, Horowitz raised the issue of "performance" standards, which left it up to laboratories to determine their own way of complying with the regulations, versus "engineering" standards, which were more specific. Frankie Trull complained that the regulations "paint a very detailed picture of laboratory animal care. They tell you point by point and that's what we object to." For instance, Part 3 of the regulations called for exercise periods for dogs in laboratories. Trull argued that there was no scientific proof that exercise contributed to the health and well being of laboratory dogs.

Aided by USDA's concilliatory attitude and Horowitz's influence at OMB, industry's objections won out. The Office of Management and Budget and the White House agreed with industry that the regulations were too specific and the proposal was shelved. But the overwhelming influence of industry on the OMB prompted Senator John Melcher, one of the authors of the Animal Welfare Act, to comment, "I have never seen anything so misinformed about the legislative process" as the White House/OMB response to those regulations.

APHIS' public show at curbing pet theft was discreetly dismissed, and industry was able to concentrate its efforts against the possibility of more potent legislation looming in the form of the Pet Theft Act.

The Pet Theft Act had been introduced in April 1988, by Kentucky Senator Wendell Ford, and in 1990 it still had not passed. Ford told legislators and the public that the express purpose of the Pet Theft Act was to "prohibit the purchase, sale or use, or transportation of stolen animals," a problem that was still rampant twenty-two years after passage of the Animal Welfare Act because of a "loophole in the law: the auction and trade day." The Pet Theft Act aimed to a vital artery of pet theft. The embattled legislation was a public relations nightmare for the medical research industry and the USDA. Its passage would confirm in the public's mind that pets were, indeed, ending up in laboratories and that USDA was not doing its job.

The requirements of the Pet Theft Act were more stringent than the USDA's proposed regulations. The Act would eliminate the use of random-source dogs and cats for biomedical research by limiting the sources of these animals to pounds in areas that, by law, already allowed sales to researchers. The bill required those pounds to hold dogs and cats a minimum of seven days before selling them to research. This would enable pets "to be reclaimed by their owners or adopted out by other individuals." Dealers could also obtain animals from individuals who raised them on their own premises, provided there was proper documentation that those dogs originated on the property. Sellers would be required to show proper ID for the dogs and cats and give detailed descriptions of the animals they bought and sold.

Failure to abide by those laws could mean severe penalties: fines of up to $5000 per dog or cat acquired illegally and permanent revocation of a license for three repeat violations.

Ford's experience with pet theft was a personal one. In 1986, a dealer had come to Kentucky to buy "stray" dogs for medical research purposes. "Unfortunately, but typically," Ford told Congress, "many of the dogs he bought were stolen—they were people's pets. Public reaction in Kentucky was loud and strong, as it has been everywhere that cherished family animals are suddenly missing in large numbers.

"I was very upset to learn of this particular incident, but it also triggered in the back of my mind the memory of the day, years ago, when my small son's prized and beloved hunting dog was thrown into a panel truck right in front of my eyes, never to be seen again. I did not know what was happening at the time, but I do now."

Ford's files bulged with documentation on pet theft—in Minnesota, Illinois, Oregon, Missouri, Maryland, California, Montana, North Carolina, Pennsylvania, Kentucky, West Virginia, Tennessee, Mississippi, Kansas, Wisconsin, and Indiana. Pet theft into research was a staggering problem. The figures presented to Congress came from a lady in Virginia who had started a citizens' network to battle nationwide pet theft. Ford told legislators, "Few people in this country know more about pet theft than Mary Warner."

In his impassioned plea, the Senator explained to opponents of the Pet Theft Act (the research industry) that limiting the source of research animals would *not* impair medical progress. The bill "does *not* come from the animal protection community; it has nothing to do with biomedical research. S. 2353 is for the people of this country who have a right to protect their pets and prevent them from being stolen. It is to assure the owners of the 54 million dogs and 56 million cats in this nation that their animals are safe."

Public outrage hurled the Pet Theft Act through the Senate. Its passage on August 10, 1988, shocked the research industry, which had considered itself inviolate.

"It's a miracle anything gets passed in animal welfare legislation," commented a Senate aide who had been involved in the furor. "Research has a lot of sway in Congress."

Some of industry's success in combatting animal welfare legislation was undoubtedly due to the incompetence of many animal interest lobbyists. Few of the multimillion-dollar animal-protection organizations had given more than a passing thought to the critical issue of stolen pets in laboratories, except during fundraising appeals. None of those well-funded groups had offered financial support to shoestring operations like Action 81 and other citizen efforts where the front-line fighting was really going on. In fact, when approached, many of these endowed groups had denied aid to local "grassroots" efforts to stop pet theft.

Among those national organizations, internal divisiveness and competitiveness over fundraising were undermining efforts at collective action and dooming animal protection legislation. While no issues attract more mail than those relating to animals, few bills fail more spectacularly. "Hammering out the language of legislation is very laborious," observed a congressional aide who worked on the Pet Theft Act. "It's not very sexy. The big groups seem to go in for the flash-in-the-pan stuff like demonstrations that show their members something tangible is happening. Well, nothing is going to happen if these animal protection interests don't have a strong, knowledgeable presence on Capitol Hill. Not all the demonstrations in the world can make up for knowing your way around the Hill."

Many animal interest lobbyists lacked surprisingly basic knowledge. One lobbyist who was trying to persuade Congress to pass the Pet Theft Act did not even know which pounds legally sold to research and which were prohibited from doing so. Some lobbyists did not know the difference between an Act and a regulation. Others did not know the sponsors of pivotal legislation. Still others predictably clung to the values of their prior positions in government. For instance, one lobbyist for a leading animal protection organization had worked as a budget officer with the Office of Management and Budget. Such an inside connection could have been a coup, but it turned out to be a disaster. While at OMB this officer had, for years, allotted USDA a minimal budget for enforcement of the Animal Welfare Act. He told another lobbyist that he thought "it was an insignificant item."

Similarly, another lobbyist for animal issues had antagonized liberal congressional allies when she previously lobbied against the Clean Air Act on behalf of her former employers, the petroleum industry, which is also a user of animals in product research.

The one shining light of unity might have been the "Tuesday Group," a monthly meeting of lobbyists from the national animal rights/protection groups. However, here too discord and ignorance prevailed. As one of the Tuesday Group's own members recently confided, "These meetings are just self-adulatory 'circle jerks.' Do you know how many congresspeople have been permanently turned off from animal issues because of that mentality? When you think of the millions of people who really care about animals and are counting on lobbyists to represent their concerns, it's tragic."

The field was wide open to industry, which took full advantage.

With a vengeance that rivaled its opposition to the Animal Welfare Act in 1966, the medical research industry mobilized for battle against the Pet Theft Act in the House of Representatives. Its strategy was orchestrated by the National Association for Biomedical Research, which had taken the industry's battered public image under its wing. The plan: to avoid the stolen pet issue by focusing on medical progress resulting from the use of dogs and cats in research.

Industry selected prestigious researchers and honed their delivery skills under this ideal diversionary umbrella. Litanies on animal rights versus human health were delivered by high-level representatives from Harvard, the Association of American Medical Colleges, the American Diabetes Association, the University of Pennsylvania, and the American Veterinary Association.

In a letter to the House, Dr. John Blinks of the Mayo Clinic denounced "this animal rights legislation." He wrote: "People steal animals only because they can sell them for a good price; the price is high only because animals are scarce; animals are scarce only because a great many pounds choose to kill them rather than to make them available for research."

What Blinks did not mention was that although Minnesota researchers had ample pound animals available to them through the state law, the Mayo Clinic and the University of Minnesota still bought from dog dealers, at least one of whose bunchers was found guilty in court of pet theft. The voluminous delivery of dogs qualified school researchers for hefty portions of the $110 million the university yearly received in federal research grant money each year.

Even Jim Glosser, despite his talk of USDA's concern about pet theft and his agency's proposals, testified *against* the Pet Theft Act. Supporting industry's position, Glosser criticized limiting dog dealers' access to auctions and the record keeping requirements imposed on research facilities. And he complained that requiring pounds to hold dogs and cats for at least seven days was impractical. A model of contradiction, Glosser said the Pet Theft Act "will hamper our ability to use the regulatory process to make changes that may enhance our enforcement efforts."

As the debate labored on, an all but forgotten issue resurfaced. Congressman George Brown wanted to know why USDA's 1987 proposed regulations for administering the amended Animal Welfare Act had still not passed. Why had they been indefinitely shelved? The cosponsor of the 1985 amendments said the delay was "absolutely reprehensible and absolutely contrary to public policy. I'm prepared to say that if I can find out if there's a culpable party here, I will try to have a negative impact on their career." Somewhere in the system, Brown said, "there's a bug that's keeping this thing from being done properly.

In December, the "bug," or rather "bugs," were isolated. The Animal Legal Defense Fund, a coalition of attorneys representing the legal issues relating to animal welfare, identified the Department of Agriculture and the Office of Management and Budget as the culprits. The Defense Fund filed suits against the Secretary of Agriculture, the Secretary of Health and Human Services, and the Director of the Office of Management and Budget charging that subversive tactics delayed passage of the proposed regulations. ALDF further charged that the Office of Management and Budget had not properly extricated itself from special-interest lobbyists.

In settlement of the case, the Justice Department set up a timetable for the passage of the regulations. The deadline for passage of Parts 1 and 2, which included regs on the acquisition of dogs and cats for research, was August 31, 1989.

Industry recognized a crisis situation and it hastily reassessed its options. Given Glosser's pro-industry stance, it would be far easier to manage USDA than a legislature stoked by citizens' wrath, as the passage of the Animal Welfare Act in 1966 had proven. In the summer of 1989, with the Justice Department breathing down its neck and House delegates pressured by their constituents to take action against pet theft, industry decided to gamble. "The research industry was gambling that the new regulations would assuage the public and Congress, and so the Pet Theft Act, which meant only more federal restrictions, would be perceived as not being needed," commented an insider on Capitol Hill.

During negotiations over the regulations, Glosser repaid industry's vote of confidence. Instead of adhering to the regulations' intent, he actually broadened researchers' autonomy. After all, his credo had always been: "Research is ultimately responsible for assuring it is in compliance with the (Animal Welfare) Act and regulations." True to form Glosser permitted facilities to apply for exemptions from the twice yearly USDA inspections. He also allowed them to determine their own method of inspections and to devise their own internal monitoring system, and he removed references in USDA's proposed regs to procedures by which lab employees can report violations of the Animal Welfare Act at their facilities. Whistleblowing never did sit well with Jim Glosser.

The record keeping requirements for dealers passed intact, as did some of the restrictions imposed on the source of dogs and cats for sale to research. The industry was not overly concerned about these concessions. After all, USDA would still be "enforcing" the rules.

Industry's gamble paid off: the Pet Theft Act died that year in the House. Its dismissal prompted Senator Ford to charge: "Those who object to my bill condone the use of stolen pets in research."

On August 31, Parts 1 and 2 of the proposed regulations, severely amended, were published in the *Federal Register*. Part 3, the treatment of lab animals, remained embattled.

In the coming years, industry's influence over government was revealed in slews of internal correspondence over Part 3.

In an April 16, 1990, letter to the Secretary of Agriculture from Alan Bromley, the Assistant to the President for Science and Technology, Bromley condemned USDA's proposed regulations as costly and unjustified, and offered a familiar litany, warning that "research capacity will be reduced in order to comply with those standards." Bromley also co-signed a similar letter with James Wyngaarden, then director of the powerful multibillion dollar National Institutes of Health. Wyngaarden would subsequently slip through the government/industry revolving door and become a liaison with the National Academy of Science. The Academy is a contract player with industry and government and, historically, an opponent of animal use reform. Wyngaarden is currently the Chief Officer for the Duke University Medical Center.

On April 17, James McRae, Jr., Acting Administrator of the Office of Management and Budget, sent back USDA's proposed regulations. He wrote to USDA's General Counsel: "It appears premature for APHIS to have submitted this draft rule." OMB needed, he said, "a more accurate understanding of the potential costs and effects on biomedical research."

In a July 17 letter, Senator Herbert Kohl wrote to the Assistant Secretary of Agriculture: "I am concerned about the process the Federal government uses to make decisions about which regulations to issue as well as what information to disseminate." Specifically, Kohl's committee questioned "the effect of OMB decision making in these areas."

USDA referred Kohl's letter to the Office of Management and Budget. OMB responded in September 1990 by claiming Executive Privilege: no answer required.

Kohl also requested documentation regarding OMB's involvement in other USDA activities. USDA responded: "On its face value it would appear to be so potentially onerous that it is not really feasible for us to assemble responsible material."

In February 1991, Part 3 of the regulations finally passed, but in the process industry had triumphed. It had successfully eliminated essential requirements of the 1985 Amendment to the Animal Welfare Act, including exercise for dogs in labs and provisions for the psychological well-being of primates. With the help of USDA, "engineering" standards which would account for the well being of lab animals were gutted in favor of "performance" standards which allowed each institution to determine for itself how it would implement the regulations. Industry's gratitude to its advocates Glosser and Arnoldi was expressed in a letter that month to the Secretary of Agriculture. Shelton E. Pinkerton, president of the American Veterinary Medical Association, wrote:

"On behalf of the American Veterinary Medical Association, I would like to compliment Dr. James Glosser, Dr. Joan Arnoldi, and members of the Animal and Plant Health Inspection Service's Regulatory Enforcement and Animal Care Branch for their exemplary performance. . . . AVMA echoes the kudos that APHIS has received from the biomedical research community for a transition from engineering standards to performance standards."

The pact between APHIS and the research industry could not have been more solidly forged.

Certain institutional record keeping would also no longer be available through the Freedom of Information Office. Among the documents protected from public scrutiny were names of clients of dog dealers and the source of research institutions' dogs and cats.

The Animal Legal Defense Fund immediately filed suit against the U.S. Department of Agriculture, the Department of Health and Human Services, and the Office of Management and Budget on grounds that those agencies had approved regulations that ran counter to the Animal Welfare Act. "Industry pressured USDA into gutting the law, even gutting the 1966 Animal Welfare Act," said Animal Legal Defense Fund attorney Valerie Stanley. "Almost anything an institution wants to do, it can do. Industry had a field day with those regulations."

Among the eleven counts the suit charged that "USDA has changed its enforcement scheme for the Animal Welfare Act," and that the "Office of Management and Budget exceeded its authority in interfering with promulgation of the regulations, contravened Congress's delegation to the USDA of rule-making authority, and violated separation of powers principles."

As of May 1992 the suit was still in litigation.

Meanwhile, a seemingly inviolate industry was savoring yet another victory. In late 1991 the Pet Theft Act passed the House, but industry had managed to obliterate its core. Because of research industry pressure on the House, its powerful connections at the Office of Management and Budget, and USDA's advocacy, auctions, the "loophole in the law" which the Pet Theft Act was intended to close, remained open to dealers and bunchers. Industry's gamble had paid off in spades.

THIRTEEN

Marshal Law

> "Dealing with these people is like dealing under wartime.
> The Marshal army has set up another code. The moral
> code that's supposed to be in place is not. Informants,
> paper trails, coded memos—it's as if we are living under
> Hitler's code."
> —*Lobbyist source commenting on Regulatory
> Enforcement and Animal Care*

May 1992
Washington, D.C.

"REAC is just a front to make it seem to the public on paper that
USDA really cares. The truth is, USDA is trying to deregulate the
industry. It's using taxpayer money to hire people who are willing to
sacrifice the public interest to give industry free rein."

For the enforcement officer talking nervously on the phone to the
author that meant he was not really supposed to do his job at Regulatory Enforcement and Animal Care. He was experienced, eager, well
trained, and committed, and he was being shut out from doing his
job protecting citizen's pets from theft. He was not supposed to enforce the law.

"Arnoldi's doing everything behind closed doors now, all the meetings. Animal Care's been given Enforcement's job, our job. We can't
even get into their computer to see what they're working on. They
want to keep everything in-house because it's such a big mess. They
are bending over backwards to excuse criminal dealers and substandard facilities. We're supposed to give out "Letters of Intent" like
candy. Go out and make inspections, and if kennels don't meet standards, all dealers have to do is say they'll send a letter promising
they'll get in shape. There's no follow-up at all. And now we've got
these Stipulation Proceedings that are letting some of the worse offenders buy their way into staying in business."

Stipulation Proceedings were the brainchild of Dr. Joan Arnoldi,
her "quid pro quo" to industry. The ostensibly cost-saving device

gave dealers an option of agreeing to pay a fine to avoid costly public hearings. Arnoldi was claiming an 80 percent collection rate from these offenders. In fact, violators were buying USDA's approval and silence—and anonymity as well, since Stipulation Proceedings were conducted in house, and out of the public eye.

The stipulation fines were often a tenth of the amount the violation would ordinarily warrant. The enforcement officer explained, "Of course Arnoldi's getting high return rates. If you were given the choice of paying a fine for a violation that, by law, should penalize you *many thousands* of dollars, wouldn't you rush to pay the smaller fine, especially if that meant staying in business?"

Did these offenders come into compliance after paying the fine?

Not according to the enforcement officer. "These dealers are still not correcting deficiencies, but they've bought themselves another year in business. Animal Care says everything is okay once the fine is paid."

The notion that USDA's new division would be devoted to enforcing the Animal Welfare Act, was quite different from what happened in reality.

"Arnoldi was never supportive of enforcement," the officer said. "I know the researchers put pressure on the agency, especially about the regulations dealing with dogs and cats. Research is a multibillion-dollar business with a lot of muscle, and government employees are not supposed to rock the boat."

His fear was palpable over the phone lines. "The brass is very paranoid about where the 'leaks' are coming from, from which sector," he said in a hushed voice.

On June 6, 1991, an internal memo entitled "Communications with the News Media" had been issued. In several sectors a magazine article written by this author was attached to the memo. That article alleged a government/research industry collusion with dog dealers. Joan Arnoldi's silencing memo warned:

> Please note that APHIS/REAC has a new policy regarding contact
> with the news media. All communication with these entities is to
> be directed to headquarters.

Why the information black-out? What was the agency trying to hide?

"They're afraid of the truth coming out that nothing is getting cleaned up," the enforcement officer explained. "People are finding out USDA has a disdain for enforcement.

"If the regs were strictly enforced, most of these dealers would be out of business. The conditions are so bad my stomach turns when I walk onto a lot of these guys' properties. Bad conditions, phoney records, dealers saying they're raising dogs when they're getting them from illegal sources. A lot of theft. That's why you've got to hammer down on the records. Records and follow-up. But that would cut the supply of legal dogs, prices would get higher, and industry would raise a stink. There's big money involved, and Arnoldi and Glosser are not going to antagonize those people. You catch hell from the Agency and dealers if you don't tell them when you're coming. Dealers won't even let you on their property if you don't call ahead. Are you going to argue with a shotgun in your face?"

Since 1988, when REAC was created, the situation had worsened. "Before REAC, my 'In' box was stacked with investigations to do, but now I've got nothing to do most of the time. I have to scrounge for work even though it's a mess out there."

Had REAC's focus ever been the Animal Welfare Act? Had its entire structure—the basis for millions in taxpayer support—been founded on deception?

Despite Glosser and Arnoldi's virulent defense of Regulatory Enforcement and Animal Care, the division's structure and track record proved more revealing. In fact, REAC's own administration would annul its charter and cripple any intended effect it might have had. Glosser had hired only forty-two inspectors to enforce the Animal Welfare Act at dealer and research facilities across the entire country. That meant, in 1989, forty-two REAC inspectors were expected to visit 1296 research facilities and 2851 sites, 4415 dealer kennels, 1504 animal exhibitors, 282 intermediate handlers, and 145 carriers.

In 1989, its first full year of operation, Regulatory Enforcement and Animal Care conducted fewer inspections of dealers and research institutions than had Animal and Plant Health Inspection Services, its predecessor in enforcing the Animal Welfare Act whose own performance had been condemned by federal watchdog agencies. In 1988, just before REAC was officially launched, and APHIS was still on the job, APHIS conducted 13,383 compliance inspections. With

REAC on the job in 1989, only 11,056 inspections were conducted. While APHIS had conducted 1913 prelicensing inspections in 1988, REAC conducted 1854 the following year.

REAC also conducted fewer inspections of animals in transit—that is, from dealer to dealer and from dealer to research. In 1989, that unit conducted 980 in-transit inspections versus the 2643 conducted by APHIS the prior year. REAC investigated fewer animal welfare complaints than did its predecessor and submitted fewer complaints to the Office of General Counsel for further investigation. When REAC inspectors did conduct inspections, their visits were preannounced (a tactic USDA still denies), giving dealers time to clean up their act. Offenders were proffered multiple warnings and "cease and desist" orders which carried no punitive action. With lax inspections, slaps on the wrists, and no follow-up, research institutions and dealers had no reason to come into compliance.

Meanwhile, the numbers of dealers increased, as did the numbers of citizen complaints and inquiries into REAC. In FY 1989, there were 64,913 inquiries about animal welfare from individual citizens and concerned groups, and 630 requests made to the Freedom of Information Act Office. Pre-REAC, in 1988, there were 52,078 complaints and 498 Freedom of Information requests.

Nonetheless, by year-end 1989, high hopes for REAC were expressed in the Secretary of Agriculture's yearly *Report to the President of the Senate and Speaker of the House*: "Now that this unit is established and fully staffed, we expect the inspection activities to increase . . . both the frequency and more importantly the quality of inspections."

As a result, REAC was allotted $7.46 million in funding for 1990, over a million dollars more than its 1989 budget. Still, REAC was hardly up to spec. While its performance had slightly improved over the prior year, the new division still was underperforming. In 1990, REAC conducted 13,050 compliance inspections of research institutions and dealers, again fewer than APHIS had conducted in 1988 when there were even fewer facilities to inspect.

In 1990, REAC conducted 175 more prelicensing inspections than it had in 1989, but only 116 more than APHIS had in 1988. There were more in-transit inspections than in 1989, but 1569 *fewer* than under APHIS in 1988! Only thirty-eight more investigations of alleged violations were conducted in 1990 than in 1988, but three hundred *fewer* violations overall were submitted to the Enforcement staff.

Meanwhile, the Enforcement staff of Regulatory Enforcement and Animal Care was not, as Arnoldi and Glosser had declared, concentrating solely on enforcing the Animal Welfare and Horse Protection acts. Enforcement still had as varied responsibilities. It was still responsible for enforcing all laws dealing with animals and plants, including Brucellosis, pet importation, and various diseases, and it was performing these critical tasks on even less money than it had been receiving prior to REAC's creation. While the budget and staff of the Animal Care division of REAC was increasing, Enforcement was still the stepchild, with fewer employees and half the funding allotted its Animal Care sister.

Yet, Enforcement's task remained as overwhelming and cumbersome. With USDA inspectors oftentimes hundreds of miles from dealer kennels, entire days would be lost in traveling. To make matters worse, when they arrived the dealers often were not present and the inspection could not be conducted. Insiders at USDA claim that officers have left blank inspection reports at kennels to be filled out by the dealer and sent back to the agency—dealers inspecting themselves.

While REAC complained about money problems, its failure was less an issue of bureaucracy and funding than of intent and philosophy. Dr. George Hoffman, a veterinary Animal Health specialist at USDA for fifteen years until his retirement in 1984, said Arnoldi's department was just what the dealers ordered. "This is just what the dealers want—no inspections, self-regulation. They (dealers) operate illegally stealing pets, pooling dogs from several sources, and hauling them in horse trailers. Dogs are dying. It's pathetic. And the USDA is still protecting the dealers."

Hoffman, an energetic man in his mid fifties, had always been outspoken on problems in USDA in the 70s and early 80s. When Glosser arrived things got out of hand, Hoffman said. A hulking six feet, his thick dark hair graying at the temples, Hoffman's presence was commanding, his voice impassioned, Old-West gritty. He had spent a lifetime trying to make the system work, and much of that time was spent battling his own colleagues at USDA.

Hoffman's responsibility as a senior official was the North-Central region: Iowa, Nebraska, North and South Dakota, Montana, Wyoming, Utah, Colorado. His jurisdiction also included what he called "the dirtiest states: Kansas and Missouri. Still are the worse."

He recalled, "Inspections have always been a joke. In training sessions, inspectors sat and made jokes. They'd make paper airplanes from the hand-outs. Why not? This is government and they know they're getting paychecks whether or not they do their job. Taxpayers should know this is where their money is going—up in paper airplanes."

But Regulatory Enforcement and Animal Care suited the biomedical research institutions just fine. Just how responsive government was to the research industry became evident in the fall of 1990, when USDA released the findings of its Stolen Dog Task Force investigation.

Confident that Glosser and Arnoldi would play ball, the medical research industry eagerly awaited the Task Force results. Glosser and Arnoldi had, after all, a commendable track record in protecting industry.

Under their hands-off policy, research facilities were not only allowed to remain in criminal violation, they were not punished for repeat violations including unsanitary conditions and failure to keep proper records of the number of dogs and cats they used in experiments and where they obtained these animals; this was in direct violation of the Animal Welfare Act.

One of the most startling examples of how the duo operated concerned a government laboratory with a history of strikingly substandard conditions. SEMA, a National Institutes of Health contract facility located in Bethesda, Maryland, had a $1.5 million annual contract to conduct research into viral disease including Simian AIDS and hepatitis. SEMA also had one of the highest mortality rates for lab animals in the country, according to People for the Ethical Treatment of Animals. When confronted with the data, SEMA denied the charge.

But the public got an inside look at SEMA on ABC's "20/20" in 1988. In addition, the BBC aired shocking videotapes acquired from the Animal Liberation Front, a radical animal rights group which had broken into SEMA. Dr. Jane Goodall, the renowned primatologist, recalled her visit to that laboratory in 1986:

I shall never forget it. Room after room was lined with small, bare cages stacked one above the other, in which monkeys circled round and round and chimps sat huddled, far gone into depression and despair. Young chimps, three or four years old, were crammed together in cages measuring 22 inches by 22 inches and only 24 inches high; they could not stand. Once they are infected with hepatitis they will be separated and placed in another cage. There they will remain, living in conditions of severe sensory deprivation, for the next several years. During that time they will become insane.

SEMA's internal reports showed chronic health and safety hazards to animals. Animal rooms were heavily "vermin-infested" and sanitation systems and liquid waste disposals were not working. In 1984, the American Association for the Accreditation of Laboratory Care (AAALAC), an independent accreditation board, found that "containment procedures and personnel protection practices in the Primate facility do not reflect a rigorous and proficient application of safety practices," and also raised concerns about the "concurrent use of carcinogens in studies conducted (there)."

In 1986, AAALAC inspectors still expressed concern about health and safety conditions at SEMA: "Experiments involving the use of hazardous material, including infectious agents as well as chemical carcinogens, are not reviewed, approved, and monitored by a safety committee." The issue was raised when a technician became infested with an internal parasite "apparently acquired from a primate experimentally injected with this parasite."

Meanwhile, hundreds of animals—some of whom had not yet been used in experiments—were dying at SEMA of unintentional causes ranging from starvation to excessive bleeding; infant monkeys also died when they were rejected by their mothers. In one plumbing accident, twenty-six animals perished, some of hyperthermia: their muscles were literally cooked.

But as far as USDA and the National Institutes of Health were concerned, SEMA had only "minor shortcomings" that required no major changes. One USDA inspector recalled, "It (SEMA) was not as bad as other places."

In 1987, SEMA officials prevented two USDA inspectors from conducting their inspection; but those inspectors decided to fight back.

On July 29, Drs. Cecilia Sanz and Janet Payeur, both high-level Animal Care veterinarians, were threatened by SEMA officials. When they tried to take photos, their camera was grabbed, and a security guard blocked their car with his truck.

The inspectors filed complaints against SEMA, and press disclosure of the episode forced USDA brass into action. The complaint was heard in-house; SEMA settled, agreeing to pay USDA a $2500 civil penalty and to "cease and desist" from further harassment and intimidation of USDA personnel. No charges concerning SEMA's violations of care and treatment of its animals were brought by USDA, but USDA announced publicly that it had "settled (the case) to enforce the humane care and treatment of animals regulated under the Animal Welfare Act."

Shortly after the SEMA incident, Dr. Sanz, who had received both her medical and veterinary degrees in her native Cuba, applied for one of five higher level positions as a USDA regional director. Her record was outstanding and her credentials impeccable, but she was rejected.

Dr. Sanz filed a complaint of sexual and racial discrimination against USDA. In 1990, when she requested her file, she was astonished to learn that shortly after the SEMA case was resolved, "Dr. Joan Arnoldi and her director of Animal Care, Dr. Dale Schwinderman, had gone to meet with SEMA. They tried to please SEMA and NIH. It was a political situation," she said.

Sanz condemned USDA's "coercive tactics." She explained, "Glosser and Arnoldi have corrupted a beautiful democracy. I left Cuba because we could not speak out, and to come to this country and see how people are intimidated and damaged—it makes me furious. I saw that Arnoldi was running USDA in Gestapo style. People are too afraid to speak out against her. They are intimidated into being quiet. But I won't keep quiet."

As of May 1992, Dr. Sanz's suit was still pending.

By contrast, at least one employee was rewarded for helping USDA save face. Glenn W. Patterson, an Animal Care specialist with the Southeast Regional Office in Tennessee, had for years been in the dog business. In 1987, the American Kennel Club suspended Patterson for one year for fraudulently registering dogs. According to an AKC spokesperson, "Fraudulent registration is one of the more serious offenses, with the same penalties as cruelty."

Patterson's side business had become too public a liability for USDA. The inspector was transferred to the Jackson, Mississippi, office of USDA where he assumed a more prestigious job as Assistant Area Vet in Charge. The agency justified Patterson's career move as "compensating" for his, ostensibly, giving up his dog business.

•   •   •

Glosser and Arnoldi were not the only ones at USDA with a track record in protecting industry. Dr. Edward Slauter, head of the USDA's South-Central region—the heart of dog-dealing country— was Arnoldi's protégé and shared her views. For ten years State Veterinarian with Missouri's Department of Agriculture and head of its Division of Animal Health, Slauter had strong political connections in a state rife with dog dealers and puppy millers. He also had a history of nonenforcement of regulations that might hamper industry.

Edward Slauter and Joan Arnoldi had met while she was still Wisconsin State Vet, but their relationship solidified in 1988. Tall, with a fair complexion and dark hair, Slauter presented a clean-cut image. He was a minister with the Re-organized Church of Jesus Christ of Latter Day Saints, a splinter group of the Mormons, a status to which he often referred during inspections. Slauter was a favorite among industry groups, including the Missouri Cattlemen's Association, which named him Man of the Year, and he was a valued member of the powerful entourage of twice-governor Christopher Bond; as such, Slauter had access to the state's Republican Party power brokers.

Slauter's connections and industry rapport must have seemed attractive to Joan Arnoldi, and when she became head of REAC she hired him as the new Chief Veterinary Medical Officer in charge of the South-Central sector. As a federal officer with REAC, Slauter was responsible for enforcing Animal Welfare regulations among animal-use industries, and dog dealers.

In choosing Slauter, Arnoldi offered a valuable "point man" in one of the dirtiest states. In keeping with his pro-industry policy while Missouri State Vet, in his new post Slauter was quite willing to leave dog dealers alone. A random look at his inspection reports revealed the same hands-off attitude that characterized the federal administration. In one case, Slauter's inspection report of a dog dealer in Frankford, Missouri, listed seventy dogs and thirty-five puppies at that kennel. The dealer had no records as to the source of these animals, but Slauter recommended no punitive action. The following year that same dealer claimed eight hundred dogs at his kennel.

Where did all these dogs come from? "Records not checked," Slauter wrote in his report.

In 1991, Slauter's questionable performance prompted Congresswoman Joan Kelly Horn to demand a USDA investigation of what a Horn aide termed Slauter's alleged "abuse of his power." These allegations included lobbying the Missouri state legislature, an action from which federal employees are prohibited. Interestingly, Slauter's lobbying efforts were directed *against* a proposed pet-protection law. He was also accused of "inadequately inspecting Class A and B facilities by announcing through phone calls his visits."

USDA found no evidence to substantiate any of Horn's claims.

Arnoldi subsequently named Slauter head of another USDA task force, this one investigating puppy mills. Its finding—minimal problems correctable through "education"—was predictable.

•   •   •

In the summer of 1990, the medical research industry was delighted to learn that their confidence had been rewarded: the Glosser/Arnoldi duo and their loyal apostles had come through on their first assignment, the million-dollar Midwest Stolen Dog Task Force Report. In October, research industry lobbyists raced up to Capitol Hill with USDA's Stolen Dog Task Force findings, parlaying those findings into a better public image to further prove that the Pet Theft Act was not needed.

Had Congress and the public been privy to the more than two hundred pages of unreleased field notes of the Task Force inspectors, they would have learned that Glosser and Arnoldi's official conclusions had no bearing on the investigators' findings.

But it did not matter that the USDA-licensed dog dealers and their bunchers were committing serious federal crimes punishable by stiff fines, suspension, license revocation, and even imprisonment. What mattered was *not* the salient evidence. What mattered was the true purpose of the Stolen Dog Task Force, and that purpose had been achieved. Glosser and Arnoldi delivered to the research industry the expected results:

> No substantive evidence was found to substantiate claims that
> licensed Class B dealers were dealing in stolen dogs or selling
> stolen dogs to research. Nearly all cases showed the dealers fol-
> lowed the federal regulations requiring them to keep accurate
> records on the origin and disposition of animals used for research.
> No evidence was found to indicate dealers were knowingly buying
> stolen animals.

As industry rejoiced in the good tidings, Glosser, Arnoldi, and their
troops circled their wagons and closed ranks. 1990 had been a rough
year for USDA, which had been under siege by watchdog agencies and
the press. The General Accounting Office and the Office of Inspector
General had cited the agency for lax enforcement of federal laws. To
add to the general mayhem, cases Arnoldi and Glosser thought were
gathering archival dust, resurfaced.

In January, the Oklahoma press announced that Governor Henry
Bellmon had determined that whisteblowers Harry Pearce, who had
since been assigned to the Stolen Dog Task Force, and Ralph Jenner
had been wronged and that the "USDA inspectors were victims of
retaliation" by their employers.

In May, a "20/20" segment proved especially embarrassing to Joan
Arnoldi. Its producers had "discovered" Alice Saffell, a USDA inspector who for years had been running her own puppy mill. Complaints
about Saffell's policies had been pouring into the agency, but Glosser
and Arnoldi had been covering for the inspector.

Unknown to the public Arnoldi had granted exemptions to her
agency's own regulations in order to give puppy millers and dog
dealers yet more leeway. In a February 23, 1990, internal memo,
Arnoldi gave Class A and Class B puppy dealers permission to transport *six- and seven-week-old* puppies within their state or to neighboring state *without health certificates*. By issuing those orders she
was circumventing REAC's published guidelines requiring that puppies be *no less* than eight weeks old and accompanied by health
certificates for shipment.

The public battering lasted through the year on news shows and
specials, including NBC's "Today Show" and "Geraldo" in May;
"Good Morning America" in July; "Inside Edition" in August; CBS's
"This Morning" in November; "Face to Face with Connie Chung" in
September.

Then in late fall, ABC News shocked the public with yet another enforcement failure of USDA. Carolina Biological Supply Company, for twenty-five years the country's leading supplier of biological materials was running a horror show. The company filled up to 3000 orders each day for preserved specimens for classroom dissection, including cats and dogs which were obtained from pounds and from USDA-licensed dealers. According to inside sources at Carolina, those dealers were stealing pets.

Individuals working undercover for People for the Ethical Treatment of Animals (PETA) submitted to the press diaries and videotapes documenting unbelievable atrocities at Carolina Biological, where live cats were cut open and pumped with formaldehyde solution while they were still moving. Some of Carolina Biological's USDA dealers reportedly paid individuals $3 to $5 an hour to collect live cats from the street by baiting them with sardines. One employee told a PETA agent that Carolina Biological "gets cats that are people's pets," and that "if people knew where we got our cats, they'd probably shut us down."

PETA also submitted the tapes to USDA, but Arnoldi had refused to watch them. Two years later, no investigation has yet been launched by USDA.

With all this bad press, confrontation over the Stolen Dog Report was just what the agency did not need.

When asked in a July 1990 phone interview with the author about the Stolen Dog Task Force Report, Jim Glosser's tolerance of inquiries was short-lived. He curtly explained, "We felt that it (pet theft) was not a problem based on that study and other indications in our inspections in the past. We did want to do this study in the Heartland, where most of the licensed dealers were located, to see if there was a real genuine concern. That an animal is stolen did not seem to be a concern. That seemed to be overstated."

Had his Task Force inspectors visited auctions?

"I am not sure of the indications of that. I simply don't know."

Did he feel proper records were kept at auctions?

Again Glosser demurred. "I am not qualified to answer that. I have to confess I don't know all the proceedings in an auction of dogs." All he knew was that the Task Force had found any illegal procurement of animals at auctions.

Had he ever been to an auction?

The USDA Administrator replied, "I have not. No." In a voice thinning with impatience, Glosser added, "The report confirms what our day-to-day inspection of records indicate. Stolen animals are not a major problem. We peruse this every day in the enforcement of the Animal Welfare Act."

Was he familiar with the names Gruenefeld, Hammond, or Huffstutler—dealers who had been investigated by his Task Force?

"No."

Did he feel that researchers should take some responsibility when it comes to investigating the sources of their animals?

Glosser said, "No, I don't—uh, we have the conditions of what constitutes to be a licensed dealer and if we find improprieties in that we terminate the license or take appropriate action."

As to why there had been only one license revocation by USDA in more than twenty years, Glosser explained, "I don't judge performance of the individual on how many convictions they get. How long is this conversation going to continue?"

Was a taped transaction proof that a dealer was buying illegally?

"I don't care about quotes," Glosser shouted. "I want facts. Documented facts. I'm not going to get into this any further." He then abruptly hung up.

Joan Arnoldi showed greater restraint when she agreed to a phone interview, two years later. "By and large we have not been able to track, find, stolen dogs," she said.

But if records are not examined at each inspection, how does an inspector know where the dogs are coming from?

"We can't check records every time," Arnoldi said.

She would not discuss the findings of falsified records at dealers, the fictitious names, or the 148 unlicensed bunchers selling to research dealers.

"I cannot deal in speculation," she explained.

Arnoldi also refused to comment on the results of USDA investigations of dealers conducted from 1987 to 1991, and she would not talk about REAC's budget. She would not answer questions about any follow-up of dealers in violation, or the number of in-transit inspections and prelicensing inspections REAC conducted in 1990. She would not comment on whether bunchers who had been offered USDA license applications by Stolen Dog Task Force crews had actually applied, or whether any punitive action had been taken against dealers who listed fictitious names. Her press officer later responded that none of the twenty-nine bunchers who had been offered license applications had applied.

Yet, as of May 1992 the majority of those bunchers were still in the dog business, and Arnoldi was insisting dealers were *not* buying at auction.

Skip Wilson, head of Regulatory Enforcement during the Stolen Dog Task Force assignment and now Arnoldi's assistant, said in a 1992 interview about the Task Force findings that he was "sick and tired of doing our job protecting animals and then hearing in the press we are a bunch of government slobs." Wilson said he had a "reputation to maintain within the agency." As far as he was concerned, there was already too much public involvement in government, as evidenced by the number of requests submitted to the Office of Freedom of Information, "which take time and don't help any animals."

As for the Task Force findings, "We didn't feel there was enough solid evidence in the report to prove pet theft. If you see a dog dealer at an auction, you've got to prove it. Who's to say they are a dealer? Who's to say he's not buying the dog for a pet?"

Only two licensed dealers were punished for inadequate records and recording false or fictitious names and vehicle numbers, among other serious federal violations. Those dealers were Don Davis of Holco, who had died in August 1990, and Bruce Barnfield, owner of BAR WAN Kennels, who was suspended in May 1991, for one year, from doing business.

BAR WAN advertised itself as "the oldest USDA Class A breeder license holder in the business." It also had a history of substandard conditions and chronic violations of the Animal Welfare Act, according to government records. While it claimed to offer "individual kennel records (date of birth and all medical data)" on its dogs, BAR WAN had failed on USDA records to identify the source of thousands of allegedly "purpose-bred" dogs on its property. Among BAR WAN's clients receiving dogs of undocumented origin: Monsanto in St. Louis, Searle Pharmaceuticals in Chicago, 3M Corporation in St. Paul, Medtronics in Minneapolis, Merrell Dow in Indianapolis and Cincinnati, Wyeth Ayerst in Philadelphia, and the University of Iowa.

As for its one-year suspension, in July 1991 BAR WAN was still advertising in the American Association for Laboratory Animal Science, which bills itself as an "educational publication." BAR WAN's ad ran near that of Purina Mill's and stanzas of the "Medical Research Fight Song" sung to the tune of the Notre Dame fight song. The research anthem begins, "Medical research that's what we do! Prevention and treatment we find the clues;" and the refrain follows, "Fight disease! Fight disease!/Research! Research!/Save our Children! Support Research!"

In a phone conversation taped in late July 1991, a caller posing as a laboratory buyer in Atlanta spoke with one of BAR WAN's employees, identified here as BW, to place an order:

LAB: I spoke with Mr. Barnfield a month ago. I want to know if he's still interested in doing business.

BW: When did you want to start?

LAB: I'm at a satellite division and he said he's got the private planes?

BW: Yup.

LAB: Do you know where he flies out of?

BW: Anywhere there's an airport, there's no problem.

LAB: Do you know off the top of your head, would he be able to sell in six weeks and deliver?

BW: How many head?

LAB: Probably about thirty a month.

BW: Thirty a month. You talkin' 'bout large dogs or what?

LAB: No, I'm talking medium.

BW: And they would all go to Atlanta?

LAB: No, some of them would go to St. Louis. I just wanted to find out what arrangement he would have about flying dogs out.

BW: What proportion would that be? About half and half?

LAB: No, it would be a whole different arrangement for Atlanta. I just wanted to check how much time he takes to fly it and how much time he guarantees for

delivery. He's still in business, isn't he? 'Cause I heard
from another dealer that he was out of business, that's
why.

BW: (laughs) He's been in business forty years. I imagine
he'll be in business forty more.

Bruce Barnfield confirmed that he was still in business in a March
1992 communication to yet another researcher made no mention of
his suspension.

USDA's Skip Wilson denied that Barnfield was in business during
his suspension, which expired May 1992.

Meanwhile, research clients of the dog dealers investigated by
the Task Force huddled around USDA's findings. Dr. Richard Fish,
Associate Director of Laboratory Animals for the University of Missouri, said, "We do not have stolen dogs here. These are not pets.
We don't see dogs come in and behave like Lassie. There may be
problems with the auctions. As to whether they were acquired from
pounds because somebody took them under false pretenses, that's a
little different."

The university bought from Holco in Arkansas, which was found
by the Task Force to be phoneying up its records including listing
one individual who was long dead; from Ray Eldridge's kennels, which
had obtained dogs from the "Free to a Good Home" ads; from Randy
Huffstutler, many of whose numerous bunchers could not be located
by USDA and who was documented by local and national news media
as buying from auctions.

University of Missouri's Department of Laboratory Animals is
headed by Dr. Ron M. McLaughlin,* who was also head of the American Association of Laboratory Animal Sciences, a laboratory animal
trade organization whose members include dog dealers. In 1989,
University of Missouri received $30.8 million in federal funding from
the National Institutes of Health for research.

Dr. Jack Hessler, McLaughlin's counterpart at Washington University in St. Louis, said he relied on the "dependability of the people
you buy from." The $1.3 billion-endowed university has been buying
for over ten years from Randy Huffstutler's Ozark Research Supplier.

---

*Coincidentally Dr. McLaughlin has the same name as the USDA official who
allegedly interfered with investigations of dealer Ervine Stebane.

Hessler, Director of Animal Laboratory Resources at the university, said adamantly, "I don't believe for a minute that dealers go out and steal dogs and sell them direct to research." Yet, in the next breath, he added, "If there's a problem the most likely avenue is through the dealers."

Was it possible he was receiving stolen dogs unknowingly?

"There is no absolute way to know. How do you know when you buy a car from a used car lot that it isn't stolen?" It all comes down, Hessler said, to "trust. If I asked them (dealers) if they go out and steal dogs, obviously they are going to say 'no' and I'll truly believe them."

• • •

Less than six months after the Stolen Dog Task Force Report was issued, Jim Glosser moved into the private sector. In June 1991 the APHIS Administrator accepted a newly created position at the University of California at Davis. There he serves a prestigious dual function as Assistant Dean of Veterinary Medicine and USDA representative in international agriculture. It is a position which ideally melds his two sponsor industries: research and livestock.

UC Davis currently has its own new status pending: that of Super Fund Toxic Site—the result of thirty years of experiments exposing beagles to radiation. According to the Environmental Protection Agency, the toxic dogs were improperly disposed of and their decomposing bodies, buried in over forty trenches, were emitting over 34,000 tons of radioactive wastes. Soil and ground water at the university is contaminated, and teenagers who twenty years before worked summer jobs in the laboratory have been diagnosed as having cancer. The disease, their physicians claim, is related to their exposure at the UC Davis beagle labs.

Shortly before Glosser's departure, USDA eased reporting regulations for its personnel. Regulatory Enforcement and Animal Care inspectors were no longer required to report on the health of the animals at USDA-licensed facilities. In fact, they were *prohibited* from doing so. This move represented a rather substantial change in policy for USDA. As John Hoyt, president of the DC-based Humane Society of the United States, wrote in a July 31, 1991, letter to Congressman Gary L. Ackerman: "Despite Federal law, (the Animal Welfare Act) APHIS has adopted a new policy that prohibits USDA veterinary and animal health inspectors to physically examine dogs at USDA-licensed facilities. This new hands-off policy negates the intent of the Animal Welfare Act . . . (and allows) kennels housing sick and diseased dogs not only to operate with impunity but with the USDA seal of approval."

By relaxing this requirement, USDA also ensured that less information on the condition of dealer dogs would be available to the public through the Freedom of Information office. The move was in keeping with agency complaints that there was too much public participation in government.

In a 1992 interview with the author, Joan Arnoldi denied that a shift in policy had taken place. In effect, nothing had really shifted, despite federal laws and congressional investigations. USDA was still industry's staunch advocate, the pact between government and special interests sealed with lucrative promises. Personal agendas and profit motives were still delivering millions of family pets into laboratories.

FOURTEEN

Hidden Agendas

Member Warning Regarding Animal Purchases: NABR
members are asked to alert their animal suppliers, particularly dog dealers, to be cautious of any unusual requests
or unfamiliar people contacting them to purchase animals. . . . One such individual (was) accompanied by a
camera crew. . . .
—*National Association for Biomedical Research*
Newsletter Update
November 1989

March 1992
Virginia

Mary Warner had had a rough night. Dwayne Reitz from the Humane
Animal League of America (HALA) had called from Hershey, Pennsyl-
vania. "Our car got blown up, Mary," he told her. "It was set on fire.
It looks like a professional job."

HALA's Oldsmobile had been packed with fliers alerting communities that pet theft was rampant. Over 380 dogs and cats had been
reported missing during the first week in March, twelve pets from
one city block alone near Harrisburg within fifteen minutes.

"They're paying kids $2 to bring them a cat, $3 for a dog. They're
paying the kids a dollar just to tell them where animals are left outside," Dwayne said.

"They" were bunchers working for a local USDA-licensed dealer.
Dwayne and his volunteers had IDed two vehicles; a white subcom-
pact and a blue pickup, "picking up dogs and cats." The bunchers
were bringing the animals to a holding area: "Sheds, buildings housing
thirty to forty dogs, and trailers packed with dogs and cats."

Dwayne managed to get one of the bunchers to talk. "The dealer
is selling to a New York lab, we don't know which one yet, but give
us time. He's saying he breeds them, but we know it's not true.
Someone already found his missing dog there, except he won't talk.
He's afraid."

The Pennsylvania state police were helping HALA, but Dwayne
suspected the local cops were getting paid off to keep quiet. "These

people mean business," he told Mary. "They sure screwed us by wrecking our car, but we're not going to stop. Nothing's going to stop us."

Normally a rousing call from one of her front-line fighters picked up Mary's spirits, but today she sounded exhausted. She had just received a copy of *Animals' Agenda Magazine*'s "Technical Report" on pet theft. It had stunned her.

Agenda, the trade publication (circulation 12,000) of the animal rights movement, was founded in 1979 in order to "network and encourage grass roots groups and activists" in a yet infant animal protection movement. Its original charter made the findings of its "Technical Report" on pet theft the more shocking.

Agenda's March 1992 report, based on what it called a "random" survey, discounted the notion of stolen pets as a source for biomedical research: "Research use is no longer a primary destination of stolen animals." It lauded the USDA's enforcement of the 1989 regulations as deterring pet theft: "The 1989 regulatory changes already appear to have caused a precipitous drop in the numbers of dogs and cats used in research." The report claimed that closing pounds to researchers actually promoted pet theft: "Demand for stolen dogs and cats was stimulated by the success of anti-pound seizure legislation." And, it utterly dismissed the dog dealer trade and its nefarious connection to research laboratories, by not mentioning it at all.

Most significantly, *Agenda* claimed that if any dogs and cats were stolen for research, the market was academia, for classroom dissection: "Classroom dissection undoubtedly accounts for far more stolen pets than biomedical research." Other markets for stolen animals were, the survey concluded, "fighting dogs, resale as pets or breeders, individual acts of cruelty, guard dogs, and for ransom."

By placing the blame onto academia and other tangential markets, *Agenda* let the biomedical research institutions off the pet-theft hook.

Agenda's Technical Report's evidence, which was collected in less than one year, dismissed more than a quarter-century of data concerning dealers and pet theft, data the Senate had relied on in passing the Pet Theft Act introduced by Senator Wendell Ford, who, in 1991, termed pet theft a "massive and heartbreaking proposition. Literally hundreds of thousands of dogs and cats are stolen from their homes each year (and sold) to research facilities."

Agenda's notion that opening up pounds to researchers would curb pet theft also defied twenty-five years of evidence accumulated by sheriffs, animal control officers, pet-lost registries, and Action 81 that pound seizure not only did *not* stop theft, but actually *contributed* to it. But *Agenda*'s findings gave credence to research industry claims.

Not only were *Agenda*'s findings suspect but the basis on which it received grant money to conduct its study may also have been questionable. The premise of the grant—the terms upon which *Agenda* received $10,000 from the Parks Foundation—was that *Agenda* would use Mary Warner's data to reach statistical conclusions about pet theft. As *Agenda*'s grant proposal stated ". . . Warner has collected and published reports of pet theft from all parts of the U.S. It should now be possible to assemble statistics from the information she has gathered which would provide a general resume of what kinds of animals are stolen, when, where, by what methods. A limited amount of similar information has been collected by the *Animals' Agenda* . . . which could also be incorporated into the study."

With that collaboration in mind, Mary and *Agenda*'s news editor, Merritt Clifton, had enthusiastically corresponded in the spring of 1991.

Mary recalled, "I was so excited, hopeful that finally all this would be public."

But once *Agenda* received its grant money, it ignored Mary's letters, and the supposed partnership dissolved. *Agenda*'s Technical Report actually attacked Mary, implying that she was manipulating her data, "tampering" with *Agenda*'s survey, and trying to "stack the deck" because of "vested interests in maintaining high theft estimates."*

Mary was utterly bewildered. Action 81's budget was a sparse $5,000 a year, largely financed by Mary.

Agenda also attacked Mary with personal slurs, telling callers to the magazine that "Mary Warner is senile, confused, and exaggerating pet theft."

In a March 1992 letter to *Agenda*, Mary's attorney warned the magazine's principals: "By implying this fraudulent activity you seriously harm her and Action 81's reputation and credibility. By implying

---

*Agenda issued two versions of the pet-theft survey results: the Technical Report and a magazine article which omitted slanderous allegations against Mary but still exonerated biomedical research as the recipient of stolen pets.

fraud, those statements are actionable under libel law." Mary asked
for a public retraction of *Agenda*'s allegations in the Technical Report.
Agenda never responded.

It would seem that *Agenda* had become an apologist for the research
industry. But why? And why was it so important to discredit Mary
Warner?

The answer lay in the genesis of the grant to fund *Agenda*'s project.
On the surface the scenario looked benign enough. But on closer look,
the money trail was littered with promissory notes to the medical
research industry.

Agenda's Pet Theft project was underwritten by the Parks Foundation, a charitable trust devoted to animal welfare. The 1992 Chairman
of the Parks Foundation, its "Chairperson of Grants," was Dr. Andrew
Rowan, Assistant Dean of New Programs at Tufts School of Veterinary
Medicine. Rowan is also chairman of one of Tufts' departments, the
Center for Animals and Public Policy, which purports to advocate
alternatives to animal experiments.

Soft-spoken, a quintessential "English gentleman" from South Af-
rica, Rowan gained renown in 1984 for his book *Of Mice Models and
Men*, a critical look at the use of animal models in research. In that
book he described biomedical research as a major financial "enterprise" that was slow to adapt to change. Rowan dismissed the industry's argument that it was "closely regulated" by calling the Animal
Welfare Act "a paper tiger," and he termed their efforts at self-regulation "inadequate." He firmly criticized USDA's poor enforcement and
the lack of standards to address painful experiments.

In the eight years since Rowan's publication, nothing had changed.
If anything, the situation had worsened. The research industry was
still self-regulated and slow to implement alternatives to animals in
experiments. There were still no standards for using anesthetics in
painful experiments involving animals. And the USDA's enforcement
of the Animal Welfare Act as well as recent regulations was still
inadequate, as revealed by the government's own internal report. The
Office of Inspector General, which reviews federal agency performance to ferret out mismanagement, fraud, and corruption, had issued its findings on USDA's Animal and Plant Health Inspection
Services branch the same month *Agenda* published its own report.
The OIG's evaluation was based on a random sample of 284 USDA-
licensed and registered dealer facilities in Illinois, Indiana, Missouri,

and Wisconsin, where about 40 percent of the country's dealers make their home. OIG concluded that the Animal and Plant Health Inspection Services, now under Jim Glosser's successor, Robert Melland, "cannot ensure the humane care and treatment of animals at all dealer facilities as required by the (Animal Welfare) Act. APHIS did not inspect dealer facilities with a reliable frequency, and it did not enforce timely correction of violations." For instance, of the dealers reviewed, 16.2 percent had had no annual inspection; 80 percent were found to be in violation but had received no follow-up inspection in the required time period; still, they were licensed despite violations including improper record keeping and identification of dogs and cats; poor sanitation; inadequate veterinary care; and substandard housing that threatened the health and well-being of the animals.

This poor report card prompted North Carolina Representative Charlie Rose, chairman of the House Agriculture Subcommittee on USDA Operations, to comment to the Associated Press, "USDA falls down on the job."

Yet, Andrew Rowan, the facilitator of *Agenda*'s survey, who had gained recognition as an outspoken critic of the USDA and research industry, and *Agenda*, an animal rights magazine, appeared to be government and industry advocates.

A look at the origin of this peculiar menage finds Rowan, in early 1991, suggesting to *Agenda*'s editor that the magazine undertake an inquiry into pet theft. Rowan advised *Agenda* to submit a grant to the Parks Foundation and he personally ushered its passage, this despite the Parks Foundation Board's reluctance to fund one of its own members. A representative of *Agenda* sits on the Parks Board and as such the magazine receives a small annuity on the $3 million interest.

While the Parks Foundation's average grant to a nonprofit is about $4000, through Rowan's efforts *Agenda* was granted $10,000.

Having succeeded in obtaining maximum funding for his pet project, Rowan became the architect of its methodology, as revealed in correspondence between *Agenda*'s editor and Mary Warner—although Rowan denied creating the methodology.

From day one, this methodology was faulty. *Agenda*'s "random" survey was, in fact, based on a selective sample of the population. It was a minute sample at that: "*Agenda*'s readers and those who sent in clippings, (Clifton's) family members, poets and writers, and (amateur) sports people, notably (those in) baseball and road racing." Since

Agenda had selected its sample, the survey was no longer random and its findings were consequently skewed.

Nonetheless, Rowan was satisfied. "I'm not surprised at the results," he said. "I'd been going around to all sorts of people about the Pet Protection Act, asking them to give me some evidence, some scale of the problem." He was directed to newspaper stories and "one or two sheriff reports (which were) just anecdotal but gave me no relative assessment of the problem."

Agenda's one-year survey provided "real statistical evidence," Rowan said. "I trust that more than Mary Warner's statistics"—a somewhat naive stand, perhaps, since Action 81's statistics had been accumulated over the course of twenty-five years from thousands of sources.

In a 1992 interview with the author, Rowan was defensive of not only the research industry but its suppliers, whom he did not believe were trafficking in stolen animals. "I don't have any evidence to say that they do [traffic]," he insisted.

Did he think USDA was doing a good job of monitoring dog dealers?

"Part of the problem is that USDA does not have enormous resources to go running around chasing after dealers. It's a time-consuming process."

Did he feel that research institutions were doing a good job of keeping records of their acquisition of dogs and cats?

"Yes," Rowan said, "I feel confident. I have no reason to question them. Those statistics I've always taken on their face value."

Rowan's support of *Agenda*'s surprising findings, his blithe acceptance of industry declarations, and his defense of USDA may reflect a conflict of interests. The major donor to his employer, Tufts Veterinary School, is Henry Foster, the founder of the largest breeder of laboratory animals in the country, Charles River, and founder of what is now known as the National Association for Biomedical Research (NABR), the lobbying entity for dog dealers and their biomedical research clients.

Both Rowan and his boss, the dean of the Vet School, Franklin Loew, underplayed Foster's role as benefactor. Dean Loew claimed, "We take a little money from them (Charles River) for scholarships." But Foster has been the chairman of the Vet School's Board of Overseers ever since the Board was created in 1988 and in that position he has generated $7 million for the school. Tufts' own newsletter lauded Foster as "the most generous supporter in the history of the school." The university's Small Animal Hospital was named for its donor, Henry Foster.

Dean Loew also overlooked mentioning his own position as a Henry and Lois Foster Professor of Comparative Medicine, a well-endowed chair paid for by the Fosters.

When Tufts' patron created the National Association for Biomedical Research, he named as its head Francine "Frankie" Trull, a twenty-nine-year old who, coincidentally, had worked for Tufts' Office of the President.

Frankie Trull recalled, "I was involved in the development of the Vet School, and in the process of putting the Vet School together I met a lot of lab animal vets, and that's how I got involved with NABR. It was very logical, actually."

Foster's view of animal use in research is not surprising, if a bit mercenary. He told the *Wall Street Transcript* in May 1979: "If you read the papers everything seems to have carcinogenic effects, but that means more animal testing, which means growth for Charles River. So you can see why we continue to be enthused and excited."

Ten years later, Charles River, which began as two rat rooms, hit sales of $85 million and was bought by Bausch and Lomb.

But the road from rat rooms to riches was not always smooth. In 1983, a University of Wisconsin cancer researcher and the university itself brought suit against Charles River for allegedly supplying the researcher and her colleagues with genetically contaminated mice, and failing to properly notify them of the problem. The result was losses in research and money, and career setbacks. While denying the charges, Charles River settled, establishing a $40,000 research fund for the plaintiffs. NIH suspended its contract with the breeder for six months when it too received tainted mice.

Tufts' patron and Andrew Rowan shared participation in another industry group: the Johns Hopkins Center for Alternatives to Animal Testing. The Center's goals are ostensibly to "develop alternatives to whole animals and disseminate scientifically correct information about alternatives." But as a beneficiary of animal-use industry money, the Center for Alternatives to Animal Testing has served the interests of members more in warding off bad publicity about animal tests than in implementing alternatives. In an interview in the Foundation for Biomedical Research newsletter, another NABR project, the Center's director, Dr. Alan Goldberg, assured FBR members that "alternatives does not mean the elimination of animals as such."

In fact, the Johns Hopkins Center for Alternatives has opened up a new, lucrative market for animal breeders and users. Both the Center and Tufts Vet School tout a potential billion-dollar market which Charles River has already cornered: transgenic animals. These are animals which are genetically manipulated: a mouse, for instance, is "developed" with a human immune system. Despite the fact that transgenic animals are still animals, the Center and Tufts promote their use as "alternatives."

Tufts, whose Dean Franklin Loew participated on numerous biotech company panels, has also ventured into this gainful market. A Tufts/industrial partnership with Genzyme Corporation has created genetically altered goats which produce human proteins in their milk in order to create a blood plasma activator used to dissolve blood clots. That biotechnically produced drug was favorably evaluated by a committee funded by the National Institutes of Health. But an international research team concluded that the substance was no more effective than its less expensive counterparts and might actually increase the risks of intracranial bleeding. It was later revealed that several members of that NIH-funded committee owned stock in Genentech, the company marketing the drug.

Ironically, the result of such academic-corporate liaisons was decried by two professors from Tufts, Sheldon Krimsky and James Ennis. In comments reprinted in the June 1992 *Government Information Insider,* a publication of OMB Watch, Krimsky observed, "A sizeable academic-industrial association will slowly change the ethos of science away from social protectionism and toward commercial protectionism."

Andrew Rowan readily admits: "I know everybody thinks I'm a show for industry. I take money from EPA (Environmental Protection Agency). I take money from industry." Bristol Meyers, one of the largest product-testing users of animals, underwrites Rowan's newsletter at Tufts. "The Center needs money," he said, "and industry is willing to give."

Among those sources is a group called Working for Animals Used in Research, Drugs and Surgery (WARDS), with whom Rowan has had a longtime association, as an editor of its publication. Rowan described WARDS as an "antivivisection group." But WARDS literature specifically states: "WARDS is not, and never has been, an antivivisectionist organization."

On behalf of industry, WARDS fought the Pet Theft Act's limitations on dog dealer sources. It also promoted the highly controversial Animal Research Facilities Protection Act of 1991, which would make it a crime for citizens to obtain information on research institutions without those facilities' permission and it would criminalize the release of certain information now available under the Freedom of Information Act. Viewed as a censoring effort, the bill has been described by the American Civil Liberties Union as "an official secrecy act for federally funded animal research . . . a sweeping assault on the public's right to know." The bill, sponsored by Texas Representative Charles Stenholm and Alabama Senator Howell Heflin, passed the Senate, and as of July 1992 was virtually assured passage in the House.

The ubiquitous Rowan also sat on several other industry boards including the Scientists Center for Animal Welfare. Their board members are "straight out of industry," Rowan conceded. "They get money from Upjohn, Johnson and Johnson. . . . They do things that respond to their membership." Rowan is also on the Board of the Public Responsibility in Medicine and Research Group (PRIM&R), which has often featured as keynote speaker, government heavyweight Dr. Frederick Goodwin. Goodwin was Administrator of the $2.7-billion federal agency, the Alcohol Drug and Mental Health Administration, and one of the country's most virulent opponents of animal research reform. In spring 1992, Goodwin came under attack for comparing "the behavior and social structure of inner-city youth with hypersexual and violent monkeys," a remark which prompted the Chairman of the Government Operations Committee in the House to call for Goodwin's resignation.

Goodwin is now head of the National Institute on Mental Health, another government branch that funds animal experiments; it is also subsidized by pharmaceutical company interests.

"He was very popular with the people" at PRIM&R, Rowan explained of his choice of Goodwin as speaker.

Rowan might have acquired his facility for keeping a firm foot in the government/industry revolving door from his mentor, Dean Loew. Dean Franklin Loew had served on National Institutes of Health committees and USDA panels and was active in the formation of the Glosser/Arnoldi division, Regulatory Enforcement and Animal

Care (REAC). Loew described REAC's beginnings as an attempt to "organize the people who do have an interest in doing the job." Among those individuals was, Loew said, Joan Arnoldi.

Loew said his experiences working closely with both Arnoldi and Glosser have been "very good. I have no reason to think she (Arnoldi) is not committed to animal welfare."

Tufts' incestuous political and industry relationships certainly edged the school into the eye of the pet-theft storm. But Tufts was in a good position to take action, it had Andrew Rowan, a conduit who amicably straddled industry and animal protection fences. He did that so well that a smiling Rowan was advertised in publications by animal interests as well as industry's NABR in which, along with AMA and government spokespersons, he *advocated* animal research.

In late 1990, while the National Association for Biomedical Research opposed the Pet Theft Act and the USDA served up its Stolen Dog Task Force findings to researchers, Dean Franklin Loew issued a directive. He explained to this author, "I told Andrew (Rowan), I said, 'help me out.' I asked him to shed some statistical light on the numbers. My instinct was that the numbers of stolen pets was not on the scale of Mary Warner's (numbers) but if someone said 'prove it, Doctor, I can't.' "

Shortly after Loew's request for data on pet theft, his protégé and minion, Andrew Rowan, approached *Agenda* with the idea for a pet-theft survey. *Agenda* must have seemed the ideal vehicle. In recent years the animal rights magazine had been plagued by internal strife. Several Board members had resigned over management and editorial disputes. One of *Agenda*'s longtime editors would later be dismissed for "physically intimidating and verbally abusing his staff." And when word got out that *Agenda* had covered up for a major scandal involving its own Chairman of the Board and one of its benefactors, the perception was that its loyalties could be bought.

In March 1990, the *Athens Daily Review*, in Texas, disclosed that the New York City-based Fund for Animals, one of the largest national animal rights groups with assets of $1.8 million and annual revenues of $1.27 million, was underwriting a market hog and cattle business on its Texas sanctuary, the Black Beauty Ranch. The founder of the Fund for Animals was Cleveland Amory, a best-selling author (*The Cat Who Came for Christmas*) who calls himself the godfather of the animal rights movement. Amory, it seemed, knew all about the livestock business since 1984. But when the press learned that Billy Saxon, the ranch manager, was using Fund vehicles and personnel for this animal slaughter business and that dead animals were left to rot near the dirty hog barns, Amory and his national director, Wayne Pacelle, issued vehement denials; this despite video documentation from WFAA-TV Dallas, newspaper photographs, and eyewitness reports.

Agenda covered up exposure of what became known as "the first scandal of the animal rights movement" by publishing articles vindicating the Fund and attempting to discredit the whistleblowers and the reporter, Charles Dukes, who broke the story in Texas.

Amory claimed to have fired Saxon in July 1990. But reports from Texas received as late as the fall of 1991 had Saxon terming himself an "advisor" to the Fund and in 1992 still running a hog business, facts misreported by *Agenda.* Meanwhile, questions have arisen about the fate of tens of thousands of animals supposedly rescued by the Fund in dramatic publicity efforts, and housed at the ranch; no records on many of those animals exist. What is known is that the ranch manager interbred domestic Yorkshire boars with the Fund's rescued wild hogs and sold those offspring for slaughter.

Since the market hog business threw into question the Fund for Animals' tax exempt status, the New York Attorney General referred the case to the office of the Texas Attorney General which took no action; such inaction, some insiders said, was due to Amory's political clout.

To further complicate the Fund's misrepresentation woes, several "grassroots" animal protection groups charged the young and ambitious Pacelle with "theft of credit," a nonprofit "no-no," that is, claiming local groups' victories as the Fund's own for fundraising purposes. *Agenda* refuted those charges on the Fund's behalf. Meanwhile Pacelle continued to give workshops on "how to handle the media."

The Fund's national director and career activist, Pacelle is Chairman of the Board of *Agenda Magazine.* The Fund is a financial supporter of *Agenda* and Amory controls much of the animal rights movement funding through his chairmanship of the New England Antivivisection Society, the multimillion-dollar funding vehicle.

As to whether money talked in *Agenda's* survey, Andrew Rowan offered: "People will pay you to do reports. The issue you have to be careful about, you have to maintain a sense of integrity about the actual analysis of the reports. Ultimately one lives in a political world."

What of the potential use of *Agenda*'s report by research lobbyists?

Dean Loew commented: "There is no question that there will be some elements of the research community that will point and gloat: 'We've been saying this all along.'"

The inference was staggering.

•   •   •

Just as dealers insulate themselves from criminal culpability by layering their buying network, the research industry camouflages its enormous influence by layering its channels of information. It is a relatively easy task. Industry fronts and foils have persuasive names like the Scientists Center for Animal Welfare, and Putting People First, a benevolent sounding group which is funded by a coalition of animal-use industries from meat to fur to pharmaceuticals. Connecticut-based "The Animal Rights Reporter" is not an animal rights publication as its name suggests, but a product of Perceptions International, an intelligence-gathering firm specializing in corporate espionage. Perceptions' clients include U.S. Surgical Corporation and other animal-use firms, which pay a high subscription fee to get "inside" information on the animal rights movement.

U.S. Surgical also paid Perceptions for its covert expertise. In a 1989 *New York Times* article the Norwalk, Connecticut, stapler manufacturer admitted it used individuals to "infiltrate animal rights organizations," a practice which it had been engaged in "since the early 1980s." Within a fourteen-month period alone, U.S. Surgical paid Perceptions $550,000 for its espionage services.

U.S. Surgical's liaison with Perceptions was unearthed in November 1989, when Fran Trutt, a self-styled animal rights activist, placed a bomb at the company's headquarters. While the national press blew up headlines like "Bow Wow Bomber" and "Puppy Love," reporters at the local *Westport News* did some digging. The result was their own headline, "U.S. Surgical Paid for Trutt's Pipe Bombs," a revelation which shattered perceptions that U.S. Surgical was simply a victim of "animal rights terrorism."

The *New York Times* and the wire services picked up the lead, the

Times reporting in 1990: "A man who has acknowledged driving an animal rights advocate armed with a bomb to the headquarters of U.S. Surgical last November now says he was a paid informant cooperating with the company and the police . . . recruited to strike up a friendship" with the accused and "to follow her activities." UPI disclosed that Perceptions International employee Mark Mead "claimed he had rehearsed Trutt's arrest with a Norwalk police sergeant and U.S. Surgical's chief of security. He talked Trutt out of planting the bomb at Hirsch's home* so all would proceed as planned."

U.S. Surgical admitted it employed Perceptions as security consultants but denied any link to the bomb or Mead, whose sister was a secretary to U.S. Surgical's security chief. The jailed Trutt pleaded guilty because, she told her lawyer, she wanted to see her dogs.

U.S. Surgical also denied affiliation with the "educational foundation," Americans for Medical Progress, which has launched a massive television and print ad campaign heralding breakthroughs in medicine resulting from animal research and decrying animal rights "terrorism." The supposed grassroots group was, in fact, founded two years ago by four U.S. Surgical employees including the firm's chairman, Leon Hirsch, its general counsel, and two spokespersons; and it is funded to the tune of $2 million by U.S. Surgical. In a July 1992 article about "seemingly independent, nonprofit groups formed to advance corporate interests," the *Wall Street Journal* dubbed Americans for Medical Progress a "public relations front," one of many whose "names invoke images of scientific credibility and concerned citizenry." The *Journal* quoted Hirsch as saying that Americans for Medical Progress "has nothing to do with us," although he conceded membership.

The American Medical Association (AMA) is currently in the throes of its own covert misrepresentation. An element of AMA's $2-million "Action Plan" to battle animal rights involved the creation of a bogus animal rights group as a diversionary tactic. The purpose of its "Foundation for Animal Health" was "to attract funding away from animal rights groups."

AMA's ruse is part of a massive public relations and advertising campaign whose aim, as described in AMA's internal documents, is to reinforce in the public mind animal research as the *only* means to

*Leon Hirsch, U.S. Surgical's CEO.

achieve medical progress, and to position critics of animal research as "anti-human, anti-science terrorists responsible for violent and illegal acts that endanger life and property."

While keeping its Action Plan strategy confined to internal documents safely out of the public eye, the AMA was attempting to defend itself from bad press arising from actions which served its own profit motive over the public health. In February 1990, the Public Health Citizen watchdog group disclosed that the AMA had voted down a Patient's Bill of Rights which would have guaranteed the right to a second opinion and information about risks and benefits of certain therapies. That same month, AMA was convicted in court of conspiring to "contain and eliminate the chiropractic profession" by violating antitrust laws. AMA also attempted to abolish a system of jury by peers for medical malpractice claims and replace it with a board comprised mainly of doctors. If the "AMA's scheme were enacted," The *Washington Post* wrote, "patients [would become] the real victims." AMA also opposed Presidential attempts to control Medicaid costs and opposed the Sheppard-Towner Act which was intended to reduce maternal and infant mortality by providing educational material to mothers. And, in 1991, the press disclosed AMA's plan to publish medical reports written by drug companies and to accept fees for services which furthered pharmaceutical interests. The plan was dropped when FDA warned it would be highly illegal.

In the arena of animal research, AMA has most effectively used scare tactics which obfuscate fact. Slogans like "Your Child or Your Dog,"—implying your child will suffer if your dog is not used in research—play well in print, TV, and promotional brochures distributed by the AMA; the "patient advocacy," industry supported Incurably Ill for Animal Research; the American Psychological Association, the American Hospital Association, the Society for Neuroscience, and the Association for Research in Vision and Ophthalmology. One of the most popular posters is offered by the Foundation for Biomedical Research (FBR), the educational arm of the National Association for Biomedical Research also presided over by Frankie Trull. Under a picture of a crazed looking "animal rights activist," the caption reads: "Thanks to animal research, they'll be able to protest 20.8 years longer."

For an industry that condemns the animal rights movement for emotional tactics, its own are surprisingly similar. In fact, researchers' use of such tactics has been described as "fraudulent and misleading" by physician coalitions which resent emotional portrayals of scientific fact. Dr. Stephen Kaufman, an ophthalmologic surgeon and chairman of the Medical Research Modernization Committee, said "the 'Your Child or Your Dog' message is a distortion of the past, present, and future utility of animal experimentation." Dr. Kaufman explained that the major advances in treating human heart and lung disease, cancer, and AIDS have come from clinical studies of patients, not through animal research.

"Given the questionable value of animal research there is no basis for claiming that there is little or no hope of future medical progress without animal experimentation. Animal experimentation is based on analogy rather than direct observation of humans and, as such, is not science at all."

When the *New England Journal of Medicine* published an inflammatory "Physicians and Animal Rights" article which portrayed critics of animal research as animal rights activists, the reaction from physicians filled several "Letter to the Editors" columns. Dr. A. Wardlaw of the National Heart and Lung Institute in London expressed the sentiments of many respondents: "Despite the millions of animals used in cancer research, over the past two decades five-year survival rates from all forms of cancer have gone up only one percent." The British scientist pointedly wrote: "I have no doubt that 90 percent of the experiments performed on animals could be eliminated without any detriment to human health." He observed that in Great Britain alternatives to animals have been widely used and medical progress has not been hampered.

The biomedical industry gets a lot of mileage from its fear campaigns, which divert public attention from the questionable value of animal research and its use of stolen pets. Animal research is big business and loss of public trust means financial disaster. Public disclosures have precipitated the loss of grant money, as Dr. Michael Carey of Louisiana State University learned. LSU, which received about $29.6 million from the National Institutes of Health in 1991, had been the home of Carey's controversial ballistics lab. The researcher received $2.1 million in taxpayer money to shoot at least seven hundred cats at close range; 70 percent of the cats were not anesthetized.

Since the purpose of the research was to study brain damage, severely wounding the cats was preferable to killing them, and large numbers of dead cats were simply trashed as useless.

An outside review committee evaluated Carey's experimental results in 1988. Its findings were expressed by brain specialist Dr. Michael Sukuff: "These jerks: they conclude that a brain-injured organism will stop breathing. I learned that in medical school in the 1950s." Sukuff was a particularly credible evaluator since he had treated ballistics head injuries in Vietnam.

Headlines in newspapers from Atlanta to Kentucky to New York and Washington read: "Shooting Cats in Head for Science Called Worthless," "Science in the Style of de Sade," "Shooting Cats: Pentagon Research Is Completely Useless." The American Medical Association defended Carey and was consequently accused of "covering up bad research." But when the General Accounting Office conducted its own investigation, the watchdog agency found Carey guilty of "high rates of failure, poor record keeping," and sloppy techniques that "could have negated experimental results."

In 1989, Congress suspended Carey's grant. Unknown to the public whose funds Carey used, many of his cats were bought from the Arkansas dealer Holco. Since Holco phonied its records there was no telling to whom those cats once belonged.

That same year, Congress severed $700,000 from a 3M Corporation bone graft project at the Letterman Army Research Institute in California when it was disclosed that the research was of questionable value, and that it would use 112 "fraudulently obtained" Greyhounds.

Letterman Institute was notorious for its mistreatment of animals. An inside source witnessed pigs shot and bled to death, mustard gas applied to the raw skin of live rabbits, and lasers directed into the eyes of restrained Rhesus monkeys. One worker called the laboratory "an awesome, terrifying place of concrete, blackened glass, and blood. (There is) an obliviousness to public scrutiny."

Greg Ludlow, Letterman's USDA-licensed dog dealer, also stocked Cedars Sinai, University of the Pacific, and University of California at Davis as well as labs in Arizona including Gore, with Greyhounds. Owners filed suit against UC Davis, former USDA Chief Jim Glosser's new employer.

In another bad publicity episode for UC Davis, a federal judge lambasted the school for filing misconduct charges against a tenured professor of veterinary ophthalmology who is also the chairman of Veterinarians for Animal Rights. U.S. District Judge Lawrence R. Karltone condemned the school for following the teacher, Dr. Ned Buyukmihci, around "like a political truth squad in the 1950s."

Public scrutiny and legislative action also resulted in the closing of a head trauma lab at the University of Cincinnati in 1988. A researcher there had been granted $1 million in public funds to crush the skulls of about 1000 cats with .22 caliber bullets. A congressional investigation was initiated by U.S. Representative Willis Gradison based on physician coalition critiques which described the experiments as "a waste of animal lives, research facilities, and money. The studies reflect unfavorably on the entire system that made the experiments possible for more than a decade—the peer review committees, the National Institutes of Health, the University of Cincinnati, and the reviewers and editors of the journals that published this work."

The University of Cincinnati obtained its cats from pounds and a licensed dealer it would not name.

The research industry's reliance on public goodwill to supply billions of tax dollars for its grants is reflected in monies channeled from one source alone. In 1991, $8.3 billion in public funds was allotted to researchers through the National Institutes of Health; in 1992, over $9 billion. Most of that money was directed toward animal experimentation in a breakdown of funds similar to the 1988 allocation, when 44 percent of the then $5.2 billion went to animal research and 23 percent went to clinical studies of human patients.

Other federal agencies that siphon many more billions in tax dollars into animal research include the Alcohol, Drug and Mental Health Administration, which conducts addiction studies; the Department of Defense, which allots about $110 million annually to test nuclear armaments, radiation, and biochemical weapons on animal subjects; the Environmental Protection Agency; the Veterans Administration, which funds all VA hospitals; the Agency for Health Care Policy and Research; the Department of Energy; and the National Aeronautic Space Industry (NASA).

Pharmaceutical companies add hundreds of millions more to the research booty at institutions, as do specific nonmedical industries funding animal research on their products, among them the cosmetic, household products, petrochemical, alcohol, automobile, tobacco, and clothing manufacturing industries.

Dr. Daniel N. Robinson, chair of Georgetown University's Department of Psychology and a member of the National Institutes of Health's grant review committee, said that denials of proposals involving animals are "extremely rare. The scientific community has a tremendously vested economic interest in research programs involving the use of animals. Their evaluation of the merits of that work should not be taken as objective."

One of the most overt displays of federal preference for animal research over treating human disease is in the area of addiction research. For instance, in 1988, Yale University researchers conducting animal-based drug addiction experiments received $3 million in government funds. That figure represented triple the state funds available for existing drug prevention programs in Connecticut and nearly quadruple the amount slated for long-term care and shelter. One grant to addict Rhesus monkeys to Valium received over $680,000—double the amount of special budgetary option funds available for drug outpatient treatment for that state.

The industry does not take kindly to such comparisons. In 1988, that analogy shut down a fourteen-year drug addiction experiment at Cornell Medical School and its researcher lost $750,000 pledged to her over the following three years. The experiments used hundreds of cats purchased from pounds and Class B dealers. The cats were fed barbiturates through tubes implanted in their stomach and their dependence/withdrawal on the drug was studied.

Cornell's and Yale's heavily funded animal addiction experiments have been widely condemned by physicians' coalitions and drug counselors, who say human drug addiction can only be studied in clinical terms. "National Institute on Drug Abuse (NIDA) is a drug industrial complex similar to the military industrial complex," said Dr. Gabriel Nahas, professor of anesthesiology at Columbia University and a consultant to the United Nations Commission on Narcotics. Nahas said NIDA, which funds addiction studies, has withheld hundreds of thousands of dollars in grants to him in retaliation for his criticism of the agency as "an old boy network dominated by entrenched interests."

As the public becomes more savvy about how its hard-earned money is being spent, biomedical research finds itself under discomforting scrutiny. The more legitimate the criticism, the more virulent industry's retaliation.

Enter the National Association for Biomedical Research (NABR—

pronounced "neighbor"). The million-dollar lobbying entity is comprised of about 350 members from academia, biomedical research, and lab animal industries, including lab animal breeders and traffickers—among them, dog dealers; NABR will not release its list of members. According to its literature, NABR's aim is to address "the flood of misinformation" issued by animal rights activists. As its 1989 Annual Report proclaimed: "In the interest of their future health and safety, the public needs to know and understand the truth. We owe them no less."

Just how this organization with such lofty goals sets about its task is somewhat contradictory. NABR, in fact, tutors its members on how to use libel suits to correct "misinformation" and fights legislation that would make researchers and dog dealers accountable to federal regulations.

A notorious 1984 libel suit provided a glimpse into the workings of the organization. The lawsuit began with a Letter to the Editor of a small medical journal. The letter, written by Dr. Shirley McGreal, chairwoman of the International Primate Protection League, criticized a plan by a multinational corporation, Immuno AG, to use captured chimpanzees for hepatitis research in Sierra Leone. Austrian-based Immuno AG is a leading manufacturer of products obtained from blood.

McGreal's letter was printed in 1983 by Dr. Jan Moor-Jankowski, the editor of the *Journal of Medical Primatology.* Immuno sued.

The National Association for Biomedical Research advocated arguments in support of member Immuno and against the eight defendants, one of whom was Dr. Moor-Jankowski, also an NABR member through his New York University affiliation. Moor-Jankowski is the longtime director of New York University's Laboratory for Experimental Medicine and Surgery in Primates, the site of about 70 percent of the nation's hepatitis research using primates. He is also a director of a World Health Collaborating Center and an internationally renowned scientist.

NABR also took an adverse stance to companies who filed briefs on Moor-Jankowski's behalf in favor of freedom of the press, including The New York Times, Time Inc., Newsday, and the Hearst Corporation, as well as renowned libel attorney Floyd Abrams.

In his book *Make No Law, New York Times* columnist Anthony Lewis termed the Immuno suit "the single most outrageous libel case—the worst abuse of the legal process." He described the suit as a classic example of those "whose sole purpose is to prevent citizens from exercising their political rights, or to punish those who have done so."

Seven years later, in a landmark decision, New York Court of Appeals Judge Judith Kaye threw out the case for "lack of substantial fact," and held for the First Amendment:

> "For many members of the public a letter to the editor may be the only available opportunity to air concerns about issues affecting them. It is often the only way to get things put right. . . . The public form function of letters to the editor is closely related in spirit to the 'marketplace of ideas.' "

But the judge issued an ominous warning. Having to litigate a libel suit all the way through trial "may be as chilling to the exercise of First Amendment freedoms as fear of the outcome of the lawsuit itself."

The chilling effect of such lawsuits may not be adverse to some interests, however. The darling of the animal-use set, Frankie Trull (who is a well-paid lobbyist for nonmedical animal industries as well) has advocated the use of libel suits and schooled her members on their value. For instance, a well-attended November 1990 NABR conference offered members a session entitled: "Defamation: How to Avoid and/or Use Libel Suits."

This kind of agitation is bound to backfire, libel-suit veteran Moor-Jankowski predicted. "For researchers really interested in doing their job, Trull's inflammatory tactics are widening the schism between animal protectionists and researchers. Such polarization ensures the need of the biomedical community for Ms. Trull's services. She is creating her own market, but she is ending up alienating biomedical researchers from an important part of the general public."

Trull's techniques have also begun to irritate members of Congress, as was evidenced during an April 7, 1992, subcommittee hearing by the House Armed Services Committee. The Committee was investigating the use of animals by the military. At that meeting, expert witnesses testified that the Pentagon kills in "excess of one half million" animals each year at an annual cost to taxpayers of "an estimated $110 million." Documentation offered by the San Francisco-based organization In Defense of Animals of materials obtained through the

Freedom of Information Act showed "a pattern of gross neglect and abuse, redundant research, and an undercurrent of disregard and contempt for civilian oversight." The federal Office of Technology Assessment reported in 1986 that "the largest user of animals experiencing pain or distress was the Department of Defense (84 percent of animals used in this category)." For instance, at the Armed Forces Radiobiological Research Center combined trauma studies required that dogs be shot and then irradiated; the dogs took a week to die and received no pain relief during that time. At the Uniformed Services University of Health and Sciences, amputation and brain injury are inflicted on kittens under four weeks old without mandatory anesthesia in an effort to determine how information is processed by the brain.

Several military experiments have already been shut down because of their questionable medical protocol or their use of stolen dogs.

In NABR's newsletter, Trull reported to her members that she had presented "a compelling statement on the necessity of laboratory animals for biomedical research" at the House hearing. The committee chairman, Congressman Ron Dellum, saw it differently.

After Trull stated that polls show a minority of the population ascribes to animal rights positions, Dellum responded that "being in the minority does not necessarily equate with being in the wrong position." Drawing an analogy to Trull's use of the term "animal rights movement," Dellum said: "I was just sitting here listening and I was trying to take out the term 'animal rights movement' and put in 'civil rights movement' or put in 'peace movement.' It was fascinating because I was part of the peace movement and part of the civil rights movement. You know what we heard? 'The majority of people do not support you. You are in the rank minority. These are a group of propagandas. These are zealots. . . .' I do not subscribe to the notion that because it is popular convention of wisdom at a given moment that has to necessarily equate with what is right." After all, Dellum said, "That is what democracy is all about."

With public alarm mounting in recent years over pet theft, NABR must have perceived another opportunity. Libel suits could work well to ward off attacks. In 1986, as the Immuno trial labored on, Trull offered advice in another libel suit. A USDA-licensed dog dealer in Michigan was suing a resident over a Letter to the Editor in a local newspaper. Rather quickly a familiar NABR scenario emerged. A parade of researchers testified on behalf of the dealer and spoke of the threat of animal activism to research. The dog dealer "is an NABR member and we provide this service to all our members," Trull explained. The dealer won his case and a considerable settlement.

Such guidance notwithstanding, Trull denied she had much of a relationship with dog dealers. "I speak to three or four now and then," she told the author in a 1991 interview.

Had she offered any advice to members about their dog dealers? "No."

In fact, in one of its *Update* publications, NABR issued a "Member Warning about Animal Purchases." NABR told its members to "alert their animal suppliers, particularly dog dealers, to be cautious of unusual requests." The press, it seemed, was onto dealers.

Trull blamed any problems with dog dealers on the USDA. "Where is the USDA?" she asked in a smoker's husky voice. "It is not like there are no regulations in place. We are always asking for more money for a better inspection program. The issue is: is the law being enforced? Our responsibility is to keep members apprised of their responsibility under the regulations (so that) they abide by them."

Once again the paper trail told a different story. NABR had mobilized to kill legislative attempts to regulate the source, documentation, handling, and care of the dogs and cats sold to research. In its own 1989 Annual Report, NABR boasted that due to its efforts those regulations were "drastically rewritten and reorganized. Many onerous record keeping and reporting provisions were deleted." Such onerous requirements included: limiting dealers' source of dogs and cats for research and mandating that pounds hold animals for seven days to give their owners time to recover them or residents time to adopt them.

In that same Annual Report, NABR apologized to its members for the fact that some "key provisions which NABR objected to remain unchanged" in the regulations. Those objectionable standards: the establishment of Review Committees to evaluate the care and use of animals in experiments; procedures to "limit the discomfort and pain to animals" during experiments; requirements that investigators consider "alternatives to procedures that may cause pain and distress"; and "written assurance that activities do not unnecessarily duplicate previous experiments."

It is common knowledge that NABR opposed the Pet Theft Act. On what basis was this opposition? Trull was asked.

She replied in an astonished voice: "We were never against it."

Fact was, Trull documented her objections to the Pet Theft Act in a May 1989 evaluation report sent to USDA: "NABR has not found any reason to believe that the legislative solutions offered by the Pet Theft Act . . . will further alleviate problems (regarding pet theft) that may exist." Senator Wendell Ford, the Act's sponsor, formally accused NABR of "misleading Congress and protecting illegal activity in its opposition to the Pet Theft Act."

How severe was the problem of pet-theft?

Trull replied: "I am unaware of any sort of broad problem of pet theft, but it is a very appealing notion for animal rights groups. I don't believe they have serious documentation. However, an institution may inadvertently receive a stolen pet. In one case they flew the animal home at their own expense a few states away."

Research facilities "absolutely have an open door policy," Trull insisted. "In most cases, they will let residents (looking for their pets), inside."

Yet in two highly publicized incidents involving stolen pets—at Cedars Sinai in LA and the Mayo Clinic in Minnesota—both facilities rejected residents' appeals until pressure from the media forced the institutions to relinquish the pets.

As for NABR members' "open door policy," Frankie Trull made the industry view known early in 1984. At a talk she gave at the University of Chicago, she advised an eager audience on how to cope with media: "by avoidance." "You don't answer these questions," she said. "In other words, they'll say 'Isn't it true that 83 percent of all animals had pain-killing drugs withheld during experimentation in your facility last year?' By the way, they know all that stuff. God bless the Freedom of Information Act, they know everything about everything."

Trull and the industry rely for their attack on animal rights on a survey conducted by the American Association of American Medical Colleges which polled 126 medical schools over a five-year period. The results, released in 1990, showed that the animal rights movement caused millions of dollars in damages to property and forced research institutions to spend $17 million a year in security to protect themselves—formidable statistics, but things were not as simple as the industry might wish. What the public did not know was that the survey-maker could not back up its own findings. "The numbers are impressive," said Dr. Douglas Kelly, the Association's vice-president for biomedical research, "but we were concerned whether they would hold up with any scientific accuracy. They probably would not."

Industry also relied on the American Association of Medical Colleges' cost estimates to delay regulations on the sources of dogs and cats and to reduce even minimum standards of animal care at research institutions. Too costly, AMA concluded.

But as far as industry and USDA were concerned, Congress and the public need never know that those surveys and polls were less than authoritative.

What does all this mean to pet-theft racketeers and their clients? Quite simply, industry strong-arming and USDA acquiescence assures easy, cheap, and virtually foolproof delivery of America's pets into the nation's laboratories.

FIFTEEN

Trials and Tribulations

"One thing I wouldn't have you do is walk my dog. Maybe
pick up its shit with your fingernails."
—*Steve Flanagan, counsel for the defense
to Ralf Jacobsen during recess*

March 26, 1991
San Fernando Courthouse, Los Angeles County
"A lynch mob atmosphere in court will favor the defense," Susan
Chasworth told her witnesses and LA activists anxious to observe the
trial. "They're going to contend their clients were victims ganged up
on by the pet owners and animal rights activists. Better stay home
until it's time for you to testify," she advised the ten victims who had
been selected. She asked the activists to wait until the day of the
verdict to come to court.

So on the first day of the trial, three years after charges had been
filed, many of those people who had invested so much time and
heartbreak in this case remained home, waiting.

The press coverage of the pretrial hearing had been extensive. But
then, Barbara Ruggiero and Ralf Jacobsen had started up yet another
"pet adoption" ring while they were out on bail. In late 1990, the
media had accompanied the LA Society for the Prevention of Cruelty
to Animals when it busted "Kimberly Lynn Christianson" and "Alexander Latham" in Bakersfield; the racketeers had selected those
names from the gravestones of babies. It was going to be difficult to
find a jury who had not read about the LA scam or of "Puppy Pavilion,"
the duo's latest scheme.

But the main obstacle counsels faced in jury selection was time.
The trial could last six months. For many prospective jurors that posed
a hardship, and they were disqualified. During the jury selection,

questions centered on pet ownership and feelings about, or participation in, animal rights groups. Most of the prospective jurors had owned a pet, found homes for pets, or put pets to sleep. Some had worked in kennels, others had bred dogs. Finding a juror who had not owned a pet was virtually impossible, but those with obvious prejudices were eliminated. For instance, one juror laughingly told the Court, "We got rid of our dog. My husband took it for a ride." Susan rejected her, a dismissal which displeased the defense.

But the defense reserved its challenges for what it maintained was the crux of this case. As Susan had predicted, the trio's attorneys wanted to try the case by arguing the merits of animal research. She had anticipated all their questions. Had the juror ever been a member of an animal rights group? Did the juror feel strongly, one way or the other, about the use of animals in medical research?

A friend of one juror belonged to an animal rights group. The daughter of another used canine cadavers in veterinary school. The twelve men and women who were finally selected had no strong feelings, one way or the other, about the use of animals in research. None belonged to animal rights groups. The selection had been as objective and representative of the community as possible. This was an intelligent, essentially neutral group of residents whose occupations ranged from probation officer to housewife and retiree.

The court was called into session and after preliminaries, Susan stepped to the podium facing Judge David Schacter. She was holding several sheets of handwritten notes to which she would occasionally refer. For day one, she wore a conservative beige suit, but her cosmopolitan style was nonetheless evident.

Judge Schacter referred to her as "Madame District Attorney."

•   •   •

For Susan, opening remarks told a story, from the beginning to the end, taking the jurors along on a journey. She spoke clearly and slowly, her concentration evident. She immediately defined the charges and attempted to diffuse the defense's strategy. "First, I would just like to say that this case is *not* about morality or the propriety of animal experimentation. This case is a theft case. These defendants are charged with stealing people's pets for the purpose of selling them to medical research. They are also charged with conspiracy to commit theft and conspiracy to commit fraud."

Susan explained, "False pretenses or fraud means that the person who is committing the theft tells the victim of the theft, makes a promise, lies about something, some misrepresentation in order to get the victim to voluntarily part with their property.

"The evidence will show that in September 1987, defendants Barbara Ruggiero and Frederick Spero went into business together under the name of Biosphere. The sole purpose of Biosphere was to sell animals to medical research laboratories for experimentation. What these two defendants decided to do was to answer ads placed in the newspaper, ads placed by people who were looking for homes for their pets. For instance: 'Black Labrador Retriever two year old, free to a good home.' The defendants told these people that they would take these animals as their own personal pets to live in their home with them. In fact, what happened to those animals is that they were sold for up to $350 a piece to medical research facilities."

Susan described for the jury how the defendants operated. "At Budget and Comfy Kennels, the pets' personal identification tags were removed and they were replaced with USDA tags. That USDA tag is required by the U.S. Department of Agriculture, which regulates the sale of animals to research facilities."

The profit margin was high. "You will learn that between December 15, 1987, and January 22, 1988, which were the dates during which Biosphere actually made sales to research facilities, Biosphere earned approximately $17,000 selling animals to research facilities."

Several jurors looked stunned. Barbara glanced at Rick and smiled.

Susan told the jury that Barbara and Rick obtained the animals themselves under false pretenses, that Barbara sold to research boarded animals while she had been paid "to keep safe and well fed." The majority of the "Free to a Good Home" animals were obtained by Ralf Jacobsen, who was "hired to get these animals that Ruggiero and Spero eventually sold."

There was a "necessary evil" in this case, Susan explained, and that was the discussion of records and documentation, since it was from records that the pets' ownership was traced. As for the defendants' motive, that was simple: greed. Their means: deception. "I'm confident that you will conclude that these defendants knowingly exploited the pet owners and their pets in order to get something for nothing. Animals that cost nothing were then sold for $350 to research facilities. And I'm also confident that you will conclude that these defendants are no better than common thieves."

Rick Spero shook his head slowly from side to side. Ralf clasped his hands in front of him and stared straight ahead. Barbara Ruggiero looked as if she could slit Susan's throat, and do so smiling.

Lewis Watnick, representing Barbara, adjusted his glasses, then rose to address the jury. It had been decided among defense counsels that he would present the opening remarks.

Watnick, in his early fifties, his sandy-colored goatee and mustache graying, began his opening remarks in a scratchy, nasal voice. "Barbara Ruggiero is charged with eleven counts of Section 487G of the Penal Code, which is a very unique and obscure Penal Code."

"Your Honor, I'm going to object," Susan said. She was about to say that the "so-called obscurity" of the code did not make it any less of a law, but Watnick hurried to finish his remark, ". . . (this is) the first time it has been used in Los Angeles."

Judge David Schacter motioned Watnick to get on with his point. The judge had a paternal manner, but it was clear his patience could easily be tried.

"It is important to remember," Watnick told the jury, "that the evidence will establish that it is not a crime to furnish animals to medical research and it is not a felony to obtain cats or monkeys or sheep or any other type of animals other than dogs within Section 487G of the Penal Code. We will prove to you that this case is not really about the taking of dogs but the use of animals for medical research."

Watnick said that the case was orchestrated and initiated by "militant animal rights interests." It was Barbara "Ruggerio" (as he frequently mispronounced her name) who was the law-abiding citizen. "She was regulated and inspected by the LA Department of Animal Regulations, and together with Rick Spero she applied for and received a U.S. Department of Agriculture B license for resale of animals for medical research, to hospitals, research facilities that also have been licensed by the Department of Agriculture. At no time did they attempt to hide the fact that they had a license and they were selling animals to approved research facilities."

Barbara and Rick went into the business of "resale of animals for medical research because a number of animals that had been boarded at her kennel were left there by the owners without paying for them and never came back. Other people would leave an animal out front, just leave it there, abandon their animals. Miss Ruggerio just could not afford to absorb the loss incurred by (these) people."

Watnick explained that his client was "victimized for filling a legitimate need for animals for medical research, that most of our recognized hospital and medical research labs have a need and most of our greatest medical advancements are——"

Susan objected.

"That is argument," the Court admonished Watnick.

Watnick emphasized the legitimacy of Biosphere, that its practices were approved by an "official in the Department of Agriculture in Washington (who) told Barbara Ruggiero and Rick Spero that 'Animals are regularly procured through the use of "Free" ads in newspapers and other periodicals, and that this was an accepted means of getting animals under Grade B licenses. . . .'" He said that veterinarians inspected these animals before and after being sold to the Veterans Administration Hospital, City of Hope, UCLA, Cedars Sinai Hospital, and Mira (*sic*) Loma Hospital.

Susan made note of UCLA. That institution had not been previously listed as an outlet for Biosphere.

Watnick portrayed Ralf as an "independent contractor, he worked for only himself," thereby underscoring the division of camps. His own client, meanwhile, was "scared and frightened by demonstrations, harassing phone calls, and break-ins" to her kennels. At no time did she have any knowledge about this obscure code section and at no time did she have any criminal intent or purpose in mind."

Watnick asked the jury not to let animal rights activists "effectively stop the sale of animals——"

"That's argument," Schacter warned.

Watnick hastily concluded his statements.

• • •

Shortly after 5:00 P.M., as Susan gathered documents in her office, the phone rang.

"How'd it go?"

"Predictable. They're making this a case against animal experimentation."

"We knew that." Norm Wegener could tell by Susan's voice that this was going to be a long, grueling haul.

"Who did you have on?"

"Bob Penia from Animal Reg. I wish they'd called one of our investigators from the police department to go into the kennels. When Reg finally went into the kennels they didn't have any idea what was going on. Many of the dogs didn't have tags on. Some of the tags were attached to the walls or cages. Animal Reg ended up destroying a lot of documentation I could have used. If only they'd recorded, been meticulous about, the identification of each dog and tag number it would have been a lot easier. More owners could have been identified."

Both counsels had ruminated over Animal Reg's handling of the Budget Boarding Kennels/Biosphere case. Enough complaints against Budget had been registered to warrant a Department investigation, well before the victims were forced to break in to claim their animals. Had Animal Reg asked the right questions, taken immediate action, might this tragedy have been averted sooner, if not altogether?

But neither attorney lingered long over speculation. There were sufficient realities to confront.

"Hank Stratton and Mike McCann testified there was no federal law in '88 to address the issue of stolen pets. There was nothing they could do, as long as the dealers kept proper records. You know, Barbara wrote down many of the owners' names. She just omitted a significant point—telling them their pets were going to medical research. USDA claims the omission is not criminal."

"This whole B dealer system stinks. Look at the scum it attracts." Wegener felt himself getting worked up. A part of him was still emotionally involved with this case.

He asked Susan about her new dog, Claude, an Irish Setter she had adopted from a "Free to a Good Home" ad. They had gone together to pick him up, both acutely aware of the irony. How easy it must have been for the defendants to walk into people's homes and take away their pets.

"You think she'll take the stand?" Wegener now asked Susan.

"I hope so," she said. "I love cross-examining liars."

April 22, 1991

The court had been in recess since March 27 because of planned vacations. Ralf remained in jail, while Barbara and Rick were out on bail. Susan had worried about whether two of her key witnesses would show up, and she was relieved to find Terry and Cindi Phillips in court on the morning the sessions resumed.

The couple had flown in from Key Largo. They had moved to the Key West islands off the coast of Florida shortly after the case broke. Terry worked at an auto store and Cindi was a department store bookkeeper.

The Phillipses felt somehow responsible for what had happened at Budget. Unknowingly, they had become party to a crime. And as did the victims, they felt guilty and angry for having been so naive.

Terry Phillips was the first to take the stand. He answered Susan's questions in a pleasant, courteous manner, describing how "Ralf would bring in an animal and say 'This is a Free to a Good Home animal.' Ralf would tell the owner of the animal about a placement program—animals going to good homes or to be in movies. Sometimes Ralf and Barbara would go out together to pick up animals. She wanted family-type dogs, medium size. She mentioned Beagles, Labs, large cats. No puppies or kittens or elderly dogs. Barbara would take her van. Ralf would go in his car. Barbara would wait around the corner and park. Ralf would pick up the animals and then meet Barbara around the corner. Barbara said it wouldn't look good for Ralf to have a car full of dogs when picking up a dog."

"Did you observe any stray dogs being brought onto the premises?" Susan asked.

"Yes. About three weeks after our employment, there was a black Lab running down the street in front of the kennel. Barbara ran out and got the dog. The dog had an ID tag with the owner's name and address. Barbara took off the collar and threw it away."

That black Lab was Candy Sheker's Pooches.

Terry told the Court that three shipments of animals went out to a "man up north" who would buy and place the animals. "The first shipment was in mid December. It was only cats, about ten of them. Barbara said they had to be in perfect health and condition. She said she wanted to impress the guy so that she would get more business. Barbara said the guy up north would call and tell her what he needed and she would get them ready. That day, Barbara left in a van, and Rick followed in his Cadillac."

A week later, five dogs and five cats, each tagged with bronze-plated USDA collars, were shipped out. At the end of December a third shipment was sent "to the guy up north. All cats, about five to eight. Barbara took them alone."

Susan questioned Terry about his and Cindi's duties at the kennels, about the secretiveness of Comfy Kennels, about his friends Bob and Benita, who had given Barbara their cats on his recommendation because the "placement program sounded so good." Terry added that his friend Jeremiah Gerbracht, an animal trainer, had also given ten of his own dogs to Barbara and Rick.

"Did any of the defendants ever say that the animals were going for research?" Susan asked.

"*Never.*"

Cindi reiterated what her husband had said. She added that she and Terry had had marital problems and that she had left after Christmas to stay at a girlfriend's house. Barbara had called her in early January and asked her to work for a day to help with a shipment. There had been ninety animals at Budget when the Phillipses quit at year-end 1987.

April 24, 1991

Before the jury entered, there was a heated debate over one of Susan's witnesses, James Manees.

"My client has informed me," Steve Flanagan, Ralf's attorney, said, "that Mr. Manees has spent time in a psychiatric hospital and he is therefore not qualified to testify."

This was news to Susan. She told the judge, "It's not a question I normally ask of witnesses, but I did give defense counsel Mr. Manees's criminal record and my notes in interviewing Mr. Manees." Manees had been convicted of two felonies: for driving without a license and escape from prison without force; his parole was up May 1991.

Watnick said, "I object to the statement by Mr. Manees in Ms. Chasworth's notes that my client offered him money to steal cages. That information is immaterial and irrelevant and prejudicial, and it should not be heard by the jury."

Judge Schacter decided otherwise. "It is prejudicial, but it is material. If you're in the business of stealing animals, you'll need a place
to put them." Schacter then instructed Susan to question Manees
regarding any psychiatric hospitalization.

Susan reported back to the Court that Manees had, in fact, been
institutionalized at Camarillo State Hospital nine years ago, when he
was fourteen or fifteen, because of a suicide attempt. "He either stole
or attempted to steal a car and he was sent to the California Youth
Authority. It was there that the suicide attempt took place."

In the hallway, Barbara talked quietly with her current fiancé, Ron
Miller, a short, muscular Army man. She was worried about Manees's
testimony. "This is trouble," she told him.

When the jury returned, James Albert Manees took the stand. Low
key, boyish looking, he explained he had been employed as Budget's
caretaker from June to September 1987. "Barbara said Budget was
for dog boarding and a cat kennel." In mid September she asked him
to pick up animals in a van or station wagon. "She said the animals
were for the movie studios. She told me to get medium-size dogs—
Shepherds and mixed breeds. She said to use the "Free" ads in the
Recycler and the pet-adoption section in the *Daily News*. I was to call
the ad and go to the house and pick up the animals. A step van would
be used; a Ford van and a station wagon. The van would be parked
out of sight of the (pet) owner. The person in the station wagon would
pick up the animal and bring it to the van."

Barbara also needed cages. "She told me that there were cages in
back of a hospital that she would like to have. She said they were
heavy. Said if I could get them, she'd pay up to $5000 for them."

Two weeks before Manees quit, Barbara told him "the animals
would be used for laboratory use. Also, she wanted the kennel immac-
ulately cleaned—she said it was going to be inspected for a license."

"Did she mention money?" Susan asked.

"Yes. I'd get $25 per animal if I picked up any. She said by selling
animals she could make good money." He never did pick up any
animals or cages.

Manees recalled seeing Ralf and Rick regularly at Budget Boarding.
He identified them both in court.

After lunch, Susan called the first of the ten victims who had been
selected to testify. The choice had been made by Norm Wegener
when the case had landed in the City Attorney's office as a misde-

meanor in March 1988. Each victim's experience demonstrated a complete, provable cycle: from the placement of the "Free" ad or the payment of boarding fees, to the tagging of the pet with USDA collars, and then to its sale to Cedars Sinai, Loma Linda Hospital, or the Veterans Hospital. These ten victims represented the 141 pet owners who had been duped by Barbara, from November 1987 to early January 1988. But there were potentially hundreds of unknown victims who thought their pets had been adopted by any one of the defendants into good, loving homes. Those dogs and cats had, in fact, been sold to research by Barbara Ruggiero through her network in California, Utah, and Mexico, where Rick planned to acquire more dogs.

During the pretrial hearing, Susan witnessed the commitment of these ten victims, their desire to see justice done. They had taken off from jobs as waitresses, carpenters, accountants, computer programmers to testify at the earlier hearings to determine whether the People had a case. They now came to the San Fernando Courthouse from all parts of California, even from a remote island off the coast of Seattle, Washington. Each was driven by a profound sense of betrayal and rage.

"These people had been lied to," Susan later explained. "They took adopting out their pet seriously. They were aggressive in their approach, and they were deceived in their own homes."

The degree to which these victims experienced the deceit and their ability to convey their sense of outrage was pivotal. Now, looking at their faces, it was clear to Susan that the years of waiting had only strengthened their resolve.

Had they known just how their pets died, these victims might not have maintained their emotional composure on the stand. Some of their dogs had died of heart attacks induced by perforation of their arteries. Some of their cats had died in hormonal tests, others in bone tests. Others had been killed in experiments which severed their nervous systems and blinded them. Still others had been the subject of classroom dissection exercises.

Mona Guinney was first to take the stand. A soft-spoken woman in her early sixties, she worked as a bookkeeper. Her daughter-in-law had accompanied her to court to offer support, and she sat anxiously in the gallery. To Susan, Mona Guinney's experience was the most disturbing, the clearest example of how cold-blooded Barbara really was. The day Barbara received her USDA license, on October 26, 1987, she immediately put a USDA tag on Flicka, her boarder of over a year. The day she cashed Mona Guinney's boarding payment for January 1988, she sold Flicka to Cedars Sinai.

Mona Guinney's testimony directly linked Barbara to selling a boarded dog to research. Nonetheless, Susan knew it would be rough proving Barbara and Rick culpable. Their defense counsels contended neither knew Ralf was lying to obtain pets. Shifting the blame to Ralf could let Barbara and Rick off the hook.

The buncher/dealer system was ideal in this kind of crisis. The set-up was tailor-made to protect dealers like Barbara and Rick from allegations of theft. Claiming ignorance of their bunchers' tactics, thousands of dealers across the country were insulated from accusations of stealing pets.

Susan showed Mrs. Guinney a photo of Flicka.

Yes, that was her dog, Mrs. Guinney said. "Flicka was a German Shepherd and Lab mix. Female. Black and white, mostly black with white legs, white breast, and white tip on the tail." A pretty dog, as Cindi Phillips had often said.

Mona Guinney explained to the jury that she had adopted Flicka when the dog was six weeks old. But eight years later she moved to an apartment where dogs were not allowed. Her son took in Flicka but could no longer care for the dog. Mona decided to board Flicka until her own situation changed. Beginning in January 1986, she paid Barbara Ruggiero $125 each month for Flicka's care. Mona, an orderly woman, had kept every canceled check but one, and the stack was taken into evidence. She also said she had visited Flicka regularly.

Susan asked: "Did you ever give Barbara Ruggiero permission to sell your dog?"

"No."

"Would you have given your dog to Barbara Ruggiero if you had known it would have been sold to Cedars Sinai?"

Mrs. Guinney's face looked drawn. "No."

Her daughter-in-law blinked back tears.

Mrs. Guinney told the Court that when she read about Budget Boarding in the newspaper she rushed to the kennel. "There were some people outside, but all the dogs were gone. I went to the East Valley Shelter to look for Flicka. She was not there. I never saw Flicka after December 1987."

The defense implied that she had asked Barbara to place Flicka, that the $125 each month was a hardship. Mrs. Guinney flatly denied the assertion.

Throughout the testimony, Barbara observed Mrs. Guinney with her usual implacable expression; occasionally she smiled. Mona Guinney might just as well have been talking about a stranger.

May 6–8, 1991

Mona Guinney concluded her testimony holding fast to her assertion that she would never have abandoned or sold Flicka. Flicka was the family dog and she had every intention of taking her back; the canceled checks were proof.

Susan called her next witness, Penny Whitman, a young woman from Sylmar. Penny's testimony was intended to link both Barbara and Rick to acquiring pets on their own, quite apart from their directive to Ralf.

In the fall of 1987, Penny was moving to a condominium and she could not take along her two dogs, Bandit, a black and white spotted Shepherd/Dalmation mix about seven years old, and Smokey, a five-year-old female Norwegian Elkhound/Shepherd mix.

Two people called responding to Penny's ad in the *Recycler*: "Two good watchdogs—free to good home." On November 1, a woman arrived at the Whitman home. Penny testified: "She introduced her husband, Rick, to me. We talked about getting married; my husband's name is Rick. Barbara said they had a nice home in Lakeview Terrace. The dogs would be like watchdogs and they would take them out walking on the trails on their ranch. . . . Barbara seemed most interested in Smokey."

Penny identified the couple as Barbara Ruggiero and Rick Spero. Rick's face darkened. Barbara smiled.

"Barbara Ruggiero did most of the talking. I told them I hoped they could take both dogs so they could be kept together. They decided to take both. Barbara said she would give them a good home. She would call me in a month and let me know how they were."

Barbara did call. "She told me the dogs were both fine and happy. She said she had given them baths."

After Animal Regulation had confiscated the animals, Penny reclaimed Bandit at the East Valley Shelter. Smokey, "the one Barbara seemed most interested in," had been sold to Loma Linda University Medical Center where she died in a classroom teaching exercise.

Like Mona Guinney, Penny Whitman firmly stated she would never have given her dogs to Barbara if she had known about the scheme.

The defense hoped to gain ground by focusing on Penny's admission that she had lied about her dogs' ages, and that she had put no conditions on giving up her dogs. Eli Guana, Rick's lawyer, asked whether she had told Rick and Barbara that they had to keep the dogs "until they (the dogs) died."

"No," Penny replied, her voice unsteady.

Watnick later asked, "Did you ask where the address was?"

"No."

"So you didn't know where your dogs were or where to find them?"

"Yes—correct."

Penny seemed shaken as she stepped down. But Susan felt Penny's testimony was strong enough to convince the jury to find Barbara and Rick guilty on Count 2: False Pretenses. Through false promises and representation, Barbara and Rick had obtained Bandit and Smokey for sale to research. There was no reason for Susan to doubt that the jury might find otherwise.

The same Saturday that Barbara and Rick had obtained the Whitman dogs, they answered a "Free" ad placed by Bill and Lorraine Greene. Susan was worried about Bill Greene's testimony. Of all the victims he seemed least emotionally involved with his dog. But the Greene case was, like Penny Whitman's, critical to connecting Barbara and Rick to criminal actions.

KK, a German Shepherd/Great Dane mix, was very large, too big for the Greenes' kids to handle; and she was a digger. The Greenes received several responses to their ad and initially gave KK to a couple who wanted a guard dog. They soon returned her: KK was "too friendly." Then Barbara and Rick came by.

Greene testified: "They said they wanted a big friendly dog for their daughter and their horse, as a companion for both. They had a ranch in Tujunga. They took her for a brief walk and said she'd be a perfect match. They seemed like a nice couple."

Rick's attorney attempted to show Greene's lack of involvement with his dog. "Isn't it true you didn't know whether your dog was a 'he' or 'she' in your prior (pretrial) testimony?" Guana asked.

"I was nervous. I believe I said 'he' and then corrected myself."

Giving KK up because she dug in the yard also did not bode well for showing Greene's attachment. Had he put any conditions on what Rick and Barbara were to do with the dog? Guana asked.

Like Penny Whitman, Greene answered that he had not. "The subject did not come up."

Susan hoped the jury would recognize the difference between the Greenes' giving KK up as a guard dog and giving her up for medical research. But if the jury felt that Greene would have given his dog to anyone for any purpose, technically no crime had been committed.

The defense continued its portrayal of Greene as an indifferent pet owner. Watnick asked Greene whether he had requested the couple's address or phone number? Greene had not.

"Hadn't Barbara kept her promise and called you?"

"Yes," Greene responded.

KK was sold to Cedars where she died in an experiment.

While Barbara and Rick were making their rounds, Ralf had also been busy. John Martin, a musician, testified that in September he had found an Afghan mix at Lake Piru where he had gone fishing. He learned from a market owner there that the dog had been running loose for days, obviously abandoned. John took him home and named him Reggie. Reggie ran away twice, and John paid $60 to Animal Reg to get him back. He tried for months to find the dog a good home with a large yard where he would be safe and happy.

John, a very likeable witness, explained that on November 1, Ralf came to his Sherman Oaks home in response to his "Free to a Good Home" ad. "He looked touched by the dog. What I mean when I say touched is that he seemed emotionally moved by the dog. He said, 'Looks like this dog found a home,' and he got down on his knees and kissed him on the muzzle. Ralf said he had a ranch in Granada Hills or Woodland Hills. I followed him to his car and told him that under *any* circumstances if it doesn't work out, I'll take the dog back. I told him I didn't want the dog to go to the pound. I said that more than once. Ralf said that wouldn't be a problem."

Reggie was sold to Cedars Sinai, but he got lucky. At the height of the press foray in February, Cedars released six dogs that were still alive; one was Reggie. Another was Sasha, a six-year-old Shepherd/ Golden Retriever whose owner, Michelle Zelman, also testified. Attractive, in her early thirties, Michelle explained that she and her husband were divorcing and needed to find a home for Sasha. "Steve Jacobs, who said he worked for the Air Force, came by. He said the dog would take care of his girlfriend while he was away." The man Michelle later identified as Ralf Jacobsen in a photo line-up at East Valley Shelter "hugged Sasha."

"Did he ever mention selling the dog to medical research or other commercial purposes?" Susan asked.

"*Never.*"

Ralf, his face pale and drawn, looked down at his clasped hands.

Elizabeth Burgos, a young mother, had been selected as the cat owner who would testify. Ellen Hickman had been anxious to appear in court, but the fate of her cats had not been clearly documented. They were not at the East Valley Shelter, and Ellen's name did not appear on Barbara's records. Most likely, Ralf had listed his own name as the cats' owner, as he had in dozens of instances. Although the Burgos case was strong, Susan had discovered that proving ownership of a cat presented a legal problem. Cats were not necessarily licensed and their vagrant habits made the issue of ownership vague. Violation of the Penal Code Section 484A, theft of a cat, was only a misdemeanor.

Elizabeth recounted a tale of deception similar to her predecessors. She had not wanted to give up Sweet Pea, a gray and white shorthaired female about two years old, and Honda, a black and brown male with white under his chin, also two. But Elizabeth's baby had been born prematurely with health problems; her doctor recommended she give up her cats.

Elizabeth turned away twenty people, and then Ralf answered her ad. "He said his name was Steve Johnson and that he was a cat lover— wanted cats that age. He petted them, held them."

"Did you give anything to Ralf Jacobsen besides the cats?" Susan asks.

"Their leashes, food, all of their cat toys. They were wearing collars and ID tags."

When Elizabeth recovered Sweet Pea from the East Valley Shelter, the cat was "very sick. Her hair had fallen out. She was skinny, shaking, had diarrhea, was throwing up. The vet said she was suffering."

Elizabeth had Sweet Pea put to sleep. Honda had died in an experiment at the Veterans Hospital.

Susan asked these questions calmly, but her witness, Elizabeth Burgos, seemed fragile, still reeling from the shattering experience.

Victim after victim offered moving testimony about the litany of lies and deception. Paul Iverson explained how "Steve Jacobs" had "knelt down and was petting PJ. The dog was kissing him on the face. He asked if the dog was good with livestock, said he had a ranch in Granada Hills."

While listening to Paul's testimony, Susan recalled that it had been the photo of PJ that made her cry. There was something about that little dog, trying to be so good while Paul took his picture. Paul's great fear, he later said, was that PJ was used in a medical school class. PJ died in just that way at Loma Linda.

The day Ralf "adopted" PJ, he also "adopted" Vanessa Haven's two cats, Ellen Hickman's two cats, Michelle Zelman's dog, James Bush's dog, and Jeffrey and Kathy Shafter's two dogs.

Blond, good-looking Jeff Shafter testified that he and his then girlfriend, Kathy Malloy, were moving into an apartment and wanted a good home for their dogs, a male Shepherd mix named Tank and a female Lab/Dingo named Domino. Someone else wanted Domino, who was pregnant, "but it was evident that 'Steve' wanted both dogs and we preferred to keep them together," Jeff explained.

Tank was sold to either Cedars Sinai or Loma Linda Medical Center. After Animal Reg's raid, the Shafters retrieved Domino and her eight puppies from East Valley Shelter. Domino was sick and all her puppies had parvovirus, a potentially fatal disease; three already had died.

"Did you use a different procedure in placing the puppies?" Susan asked.

"Yes. Before we let people take a puppy we asked to see their driver's license and got their phone number. We offered to take the dog back if the circumstances didn't work out. We checked up on the people afterward."

Elaine Goodfriend, a housewife with several children, testified, "I will never forget this as long as I live. He, 'Steve Jacobs,' got down on one knee, put his arms out, and the dog kissed his face."

Elaine recalled walking her dog to Ralf's car and noticing another dog in the back seat. "A large black dog, an Afghan mix with long hair." That dog was John Martin's Reggie.

Barbara observed with detachment the parade of distraught victims. When they looked at her, she smiled. When they identified her, her smile broadened. She appeared almost serene. By contrast, Rick was tense, his eyes ringed with dark circles, and he whispered frequently to his counsel, Eli Guana. Ralf was pale and seemed to have lost weight.

On May 7, Chuck Ransdell was called by the prosecution. Chuck had been waiting three years for this day. He confidently strode on long, lean legs to the witness stand.

The loss of Ammo hit the Ransdell family hard. His wife, Leslie, could not concentrate at work for months; she could not stop crying and had trouble sleeping. Their son, Travis, kept asking about Ammo.

What was he supposed to say to a seven-year-old?

When Chuck learned about the scam, he immediately called Cedars, but officials there denied accepting Ammo. He called Barbara's lawyer, at that time, Hugh Seigman. Seigman told Chuck that Ammo had "*definitely not* been sold to a research facility."

"You think if someone's going to cheat they'll do it on their income tax, not do something like this," Chuck later said of Barbara's scam. "And Barbara knew Ammo for years, ever since he was a puppy. She had her own dogs, for God's sake."

When it was disclosed through Biosphere's records that Ammo had, indeed, been sold to Cedars, Chuck sent a letter to the institution threatening to sue for damages. Money was not the issue. He just needed to take some action, for the sake of his family, for Ammo.

At the pretrial hearing, Chuck was remanded into custody for contempt of court when the defense accused him of extortion.

"Are you nuts!" Chuck shouted. "They're the ones that committed the crime!" He was released pending an attorney's letter explaining the intent of his communication with Cedars, and the "extortion" matter was dismissed.

Shortly after, Chuck joined an animal rights demonstration at Cedars Sinai.

"I never thought I'd be in one of those demonstrations, I'm not a joiner, I don't get involved," Chuck later said. "But I felt I had to be there, for what Ammo meant to my family. There I was watching these doctors hanging out Cedars' windows laughing. I sure got a different view of what's going on there."

Now, on the witness stand, Chuck stared at Barbara's impassive face. His own reflected rage and sorrow and his lost faith in the judicial system.

Susan began her direct examination. "At some point did you determine you had to find a new home for your dog?" She asked Chuck.

"Yes, our dog needed more room. He was getting out of the yard, eating the fence." Chuck explained that he had tried several animal adoption services but was told that placing a four-year-old dog was difficult. They tried training Ammo, but he was wild, a real untamed character.

During a pretrial hearing Chuck told the Court that in July 1987 he had dropped Ammo off for his usual summer boarding while the family went on vacation. But this time he asked Barbara "if she knew of someone who would place my dog. Barbara said she had a friend in Altadena or Pasadena and for a $35 fee she would place dogs. I sent her a check and a detailed statement of the traits of my dog . . . I told her Ammo needed a large yard because he was so big."

Chuck's canceled check for $35 was taken into evidence as proof of the embezzlement.

When Chuck returned home from vacation, he called Barbara. He explained, "She said the dog was with a family in Canyon country and that they had two boys. I told her that if there was any problem with Ammo that she should call me—that I didn't want the dog to go to the pound. Even if it's a year later, I told her to call me and I'd pick up the dog—I'd eat the $35."

Had Barbara ever mentioned Ammo's going to medical research?

"No. I would never have given him up then. I didn't even want him going to the pound."

Chuck did check up on Ammo. "I called her before the holiday—between Thanksgiving and Christmas. My son wanted to see the dog. Barbara said he was fine, was still with the family on two acres and had settled in. I thought it would be upsetting to my son to see the dog again so I let it go. I again told Barbara that if there were any problems I wanted the dog back."

Chuck was still blaming himself for not going to see Ammo. When he learned of the scam in January 1988, he called Barbara's attorney. "He told me to fuck off," Chuck said.

Ammo was sold to Cedars, where he died in an experiment.

Watnick tried to establish that there was no way to determine that Ammo was *not* adopted before being sold to Cedars.

"From July 10 until January of 1988 you had no idea of the whereabouts of your dog?" he asked Chuck.

"Yes. But I had an idea that the dog had been placed."

Had Barbara intended all along to sell Ammo for medical research? Or had she sold Ammo after this alleged adoption fell through?

"Intent" was the key to making Barbara criminally culpable. Susan hoped the jury would see through the smokescreen of this supposed adoption to a family whose name Barbara told the Court she did not even know.

On May 8, Norman Flint was called to the stand. As Susan watched him approach, she recalled their first meeting in her office three years earlier, when he struggled against his tears. That day Norman had told her, "I don't want to start bawling when I get up there." She said that he would do fine, that it was all right if people saw how much he hurt.

Norman had left his home on Orcas Island, off the coast of Seattle, Washington, at 4:30 a.m. and flown a four-seater to Friday Harbor. There he had taken a twelve-seater to Seattle, 70 miles east, then jetted from Seattle to Burbank Airport outside LA. The distance had not mattered; nothing could have stopped Norman from testifying at the trial.

Norman had learned of the scam through his sister. She had called him in late January from the East Valley Shelter where, coincidentally, she had gone looking for her cat which had wandered. She heard from volunteers there about the theft ring. Through a network of victims, Wiggles came back to him. Candy Sheker had found Wiggles at East Valley Shelter when Animal Reg confiscated Budget Boarding's animals. Candy had not wanted to leave the starved-looking dog behind and so had claimed as her own not one young black Lab but two, her own Pooches and Wiggles—a poor reflection on Animal Reg's monitoring of dogs that were released, but a generous move on Candy's part, one that brought Wiggles and Norman together. But that reunion was circuitous. Candy hooked into a network formed by victims who were trying to find their pets—dogs and cats they thought had been adopted into good homes by the defendants. Wiggles was then claimed by Denise Tracey, whose husband was convinced the skinny dog was their Duke. He told his wife, "After what Duke went through, sure he'd look different." But Denise was not convinced. She phoned the victims' network, looking not only for her Duke but for the real owner of this Lab. The day Norman Flint pulled up to her house and Wiggles went bounding out to meet him, she knew the dog was finally going home.

When he learned about the fate of Bear, Norman could not have gotten far enough from LA. Life on the remote northwest island had healed him of the pain of losing his buddy. On clear winter nights he could see the Aurora Borealis, the dazzling northern lights, and when the weather was milder he took Fred, Cody, and Wiggles swimming. The building company he started was doing well. He was beginning to trust in life again.

But during the journey back to LA, rage and sadness flooded his heart. He recalled the pretrial hearing, when the defense cracked jokes. He had told them, "I don't think there's anything about this that's funny." What kind of people were they, anyway?

At that initial hearing, Watnick had tried intimidating him. "He used a lot of big words," Norman later recalled. "He knew the right words, but I had the truth."

Now, three years later, Norman Flint took the witness stand and locked eyes with Ralf Jacobsen. A sobering thought hit him: he could be the one on trial now for killing that son of a bitch.

In a husky, emotional voice, Norman described how he had rejected twenty people until "Mike Rogers, who said he was a lawyer," came along and promised Wiggles and Bear "a big house and yard, a family, a good home."

Later, Norman sought out Ralf. "I told him I'd like to beat the shit out of him but that I was not going to get in trouble for doing something wrong. I told him I'd see him in court."

"Were you upset?" Susan asked.

"Yes. I thought my dogs were dead. He had lied to me—shook my hand and lied."

Norman explained that he returned to Ralf's house. "He said he wouldn't come out. I took a picture of the house, the address, and the car he took my dogs away in."

Those photos were taken into evidence. Norman said he called Cedars, talked to a Dr. Young who said his dogs were not there. He called the VA Hospital and got no answers. "I went to the East Valley Shelter six or eight times and went to the West Valley Shelter two times. I never saw Bear again."

Bear was killed in a heart attack experiment at Cedars.

Watnick approached for cross-examination and Norman visibly tensed. The lawyer challenged Norman's memory: was it really Barbara who had called about his dogs?

The prosecution and defense argued about his prior testimony until Norman slammed his fist down on the stand.

"Look. They called—he, she, husband, wife, who cares! I know that Barbara called and then Ralf came over. He said his wife's name was Barbara."

Watnick suggested that Norman was mistaken. Perhaps Mr. Flint did not understand English.

Norman growled, "If you spoke it clear, man, I'd understand it."

The jury laughed and there was a slight relief of tension in the courtroom. Bear's owner, a big bear of a man, heaved himself off the stand. The courtroom seemed to hold its collective breath as Norman shot a scathing look at Ralf, then walked out.

In the hallway, Norman Flint sat on a bench and covered his face with his big, calloused hands.

SIXTEEN

Co-conspirators

"The whole thing is a cover-up. . . . The research facilities, why weren't they brought up on charges? They were
paying $350 a dog, for God's sake! That's our tax money
and what do we pay it for—for them to steal our dogs?"
—*Candy Sheker, victim*

May 27–28, 1991
San Fernando Courthouse, Los Angeles County
Susan Chasworth spent the evening of May 27 conferring with Gary
Olsen from Animal Regulation. Olsen was scheduled to testify in the
morning, although the bulk of his testimony would come later, in
June. He told Susan that he was still stunned that a dog dealer had
been operating right in Los Angeles, in his own jurisdiction. He
appeared genuine, a conscientious, if not a bit cautious, public ser-
vant. Susan predicted he would do well on the stand.

At the close of their meeting, they speculated about the defendants.
Would they take the stand?

Susan believed that Rick, the glib business manager, would testify.
She couldn't imagine Ralf speaking on his own behalf.

"I don't know how in the world Ralf is going to explain away the
evidence we have against him. What could he possibly say in his
defense—that he was blinded by love of Barbara? I think he'd be in
a very awkward place saying these animals were being obtained for
any other than commercial purposes."

Did she think Barbara would testify, Olsen asked her.

"For someone who wants to be perceived as 'respectable,' I think
she would have a hard time explaining herself to the court," Susan
said.

Would this expert manipulator try?

"Barbara Ruggiero might just have the arrogance to think she could
pull it off," Susan told Olsen.

Olsen handled himself well on the stand. He had the bearing and clarity of an athlete, and an appealing manner. Pensive and thoughtful, he related in detail the content of his meetings with Barbara and Rick, and their then-attorney Hugh Seigman.

"I asked them if they sold to medical research. Both responded 'yes.'"

Barbara had told him that Ralf answered "Free to a Good Home" ads and brought the animals to Budget Boarding.

"She said he was not to use proper names in acquiring the animals. Not to lie, but not to make promises he could not keep. He was not to tell the pet owners where the animals were going. Mr. Spero instructed (Ralf) not to say anything about medical research, and not to imply anything of a contractual nature to the pet owners. They said they had a USDA license."

Eli Guana grilled Olsen on the legality of Budget's operation. Didn't they satisfy Department of Animal Regulation?

"Yes."

Did their records indicate criminal activity?

"No."

Wasn't the problem really that animal rights groups were pressuring the Department?

Olsen bristled. The question affronted the integrity of the career officer. "No," he replied tersely.

The months on trial had taken their toll on Rick and Ralf, but Barbara looked remarkably well. Her hair was freshly washed each day and her clothes were clean and pressed. She flirted with the security guard while counsels argued among themselves, and appeared incredulous that she was on trial at all.

During a recess Barbara offered her perspective on the trial to the author. If anyone was to blame, she said, it was the victims.

"I was providing a service to responsible institutions," she explained. "But people were not honest about why they gave up their dogs—their age, pregnancy, they were diggers, barkers. So who is the liar?"

The entourage that surrounded her—Rick Spero; her current boyfriend, Ron Miller; her sister, Jo; and her attorney Watnick nodded in agreement. "USDA says 'random source,' that means 'Free to a Good Home' ads, up until recently. I had no knowledge what this gentlemen Jacobsen said. I wasn't there, I told him not to lie."

But had she told Ralf to say the animals were going for medical research?

"I've given my own dogs for research. I tried training one with an electric collar, but he'd kill chickens, dig. So rather than put my problems on other people, on society, I gave him to a place where he'd do some good."

"The problem is with the USDA," Spero said. "The law. Random source. What does that mean? Dogs from all sources."

Ron Miller complained about the "hatchet job" by the press. "Barbara should not talk to anyone unless she is compensated."

Her sister, Jo, agreed. "She's been through enough. She should get something out of this."

Barbara smiled shyly. "Yes," she sighed. "These three years have been a nightmare. I've gotten notes, death threats, photographers watching what I'm doing. Harassment. I've been trying to live my life as best I can. I've gone from valedictorian to a felony trial. The whole thing is distorted, blown up."

Of course the whole thing was ridiculous, her defense attorney said. "No promises were made. The question is whether the dogs were obtained under false pretenses, but the prosecution is trying to make this a case about whether animal rights is right. We're trying to bring this back on track."

Barbara returned to court, banked by her devotees. As she passed a security guard, she smiled at him.

The next witness was Dr. Charles Kean. Kean, the director of Loma Linda's Animal Care facility, was responsible for the veterinary care and acquisition of its research animals. Loma Linda Medical School was the focus of controversy in the mid 80s when the heart of a baboon was transplanted into a human known as "Baby Fae." The surgeon, Dr. John Bailer, had been experimenting with trans-species transplants for years, with little success. The failed baboon-to-baby transplant was widely condemned by medical professionals and ethicists.

Kean, an attractive man in his mid forties with salt and pepper hair and mustache, and thick eyebrows had the calm demeanor of an academician. Under Susan's questioning, Kean explained that he had learned of Biosphere through its circular. He placed an order for eighteen animals in mid December 1986. Exhibit 57, a purchase order form, was taken into evidence.

Kean explained to the Court, "The purpose of a purchase order is that we will reserve funds to cover payment and that, once delivery of the product is received, we send an invoice to the accounting department and they pay it."

The "product" Kean referred to was eighteen dogs, weighing from twenty to forty pounds. Loma Linda had paid Biosphere $4770 for a January 6, 1988, delivery.

Susan displayed USDA tag #17. Kean identified the number on a Loma Linda order form.

And USDA tag #51?

"Yes, on the second page."

"Number 47?"

Also on that page.

"Number 38?"

Kean identified that number, as well.

USDA tag #17 was Penny Whitman's Smokey. Tag #47, Jean Virden's Siberian Husky. USDA tag #51 was Paul Iverson's PJ, who had died in the kind of "dog lab" exercise which has been eliminated by many medical schools in favor of nonanimal alternatives.

USDA tag #38 was Elaine Goodfriend Adler's Dojjy.

Kean told the jury that by January 12, all the dogs had been "euthanized."

Rick's attorney attempted to channel the discussion to Loma Linda's need for animals. "Why do you use animals?" Guana asked Kean.

"Irrelevant," Susan objected.

Judge Schacter said loudly, "Sustained."

"Could you function as a medical facility but for the use of animals?" Guana asked.

Before Susan could object the Judge retorted, as sharply, "Sustained."

Under cross-examination by Barbara's lawyer, Kean said he purchased animals from "various dealers." He had contacted Dr. John Young at Cedars to verify Biosphere's credentials.

Next, Dr. John David Young approached the witness stand. He had the self-confident air of a man accustomed to professional respect and attention. Tanned, tall, and blond, in his mid thirties, Young was the laboratory animal veterinarian employed by both the Veterans Administration Hospital in Sepulveda and Cedars Sinai Medical Center. For the past seven years he had purchased laboratory animals for both institutions. Young was also a spokesperson for the California Biomedical Association, promoting the use of animals in research.

Under direct examination by Susan, Young told the Court that he had made an on-site visit to Biosphere on November 27, 1987, after receiving a flier advertising Biosphere's USDA-approved services. The jury listened attentively as Young described his visit. Conditions there were "sanitary, good. I was basically inspecting the facility like the USDA would inspect my facility. I asked for one extra piece of verification—proof of the five-day holding period and the source of the animals."

The mandatory five-day waiting period was ostensibly to ensure the animal was not harboring infection. In reality however, as Barbara's victims had testified, their dogs and cats entered healthy but those that returned home were ill or near death.

The five-day waiting period was also an attempt to give pet owners time to claim their lost or stolen pet from a pound or dealer. Here too the reality was quite different. Unscrupulous pound managers across the country—many of whom moonlighted as dog dealers and bunchers—quickly turned over premium research dogs like KK, Bear, and Flicka to researchers. And since USDA did not enforce the holding-period regulation, dealers rarely complied, particularly when the source of their animals was local. Local residents finding their "lost" pets at dealers and laboratories was clearly problematic.

Young testified that he received a two-page document signed by Rick Spero indicating that they were aware of the holding period, and that they "didn't limit their source of animals to the state of California." That letter was introduced as evidence. Young said that Spero told him "some animals were raised on the premises."

Susan introduced into evidence several other USDA forms, dated December 15, 1987, and January 7 and 22, 1988, shipment days from Biosphere to the VA Hospital; a total of twenty-eight cats.

Did the USDA tags identify each cat? Susan asked.

"Yes."

Tag #92, for instance—sold to the VA by Biosphere?

"Yes, on December 15, 1987. Male tabby—7 pounds, received 1/7/88, euthanized 1/13/88."

The male tabby #92 was Elizabeth Burgos's Honda.

Young testified that on January 4, 1988, Cedars Sinai received a shipment of fifteen dogs from Biosphere. They weighed between forty and seventy-five pounds; cost, $350 each dog. On January 4, 1988, the amount paid by Cedars Sinai to Biosphere was $5,250.00—for one shipment. On January 21, Cedars received another shipment of fifteen dogs. Cedars paid Biosphere $4,908.75 for that shipment, which included a refund of $341.

Were the dogs identified by tag numbers?

"Yes."

Susan displayed USDA tag #49 for Young's inspection.

"Yes," he said, recognizing the tag.

USDA tag #49 was Kathy Malloy Shafter's young Shepherd, Tank. And tag #81?

"Yes."

USDA tag #81 was Chuck Ransdell's Ammo.

Barbara listened with detached interest as Young identified USDA tags #46, #37, #21, #14, #93: Al Jenkins's Rhodesian Ridgeback, John Martin's Reggie, Bill Greene's KK. USDA tag #14 was Flicka. USDA tag #93 was Norman Flint's Bear. Wiggles, USDA tag #94, had been scheduled for the next shipment.

Young told the jury that the dogs were used for "surgical procedures, cardiovascular research," but he did not explain what that research involved. These procedures included induced heart attacks and other painful invasive experiments from which the animal subjects did not recover.

Young did not tell the Court that Cedars refused to relinquish the six dogs still alive when the case broke. Only when the LA press pounced did Cedars release photos of the six. Their owners had been given forty-eight hours to claim their dogs, and then experiments would commence. Reggie was one of the lucky ones. So were the dogs belonging to Al Jenkins, Michelle Zelman, Georgia Rogers and, Kathy Shafter as well as another Shepherd mix whose owner was not found, but was adopted by another victim.

Young told the jury that on January 26, 1988, Rick Spero had called him. Rick told Young that he had a problem with his kennel. "He said an animal rights group attacked his facility and stole animals." Rick informed him that "Free to Good Home" animals may have been inadvertently sold to Cedars and the VA, and that he would give Young a list of tag numbers of animals that may have been obtained "improperly."

Was he (Young) able to produce all the animals Spero identified? Susan asked.

"I don't know for sure. I don't have documentation."

The aloof Dr. Young conveyed the impression that "Free to a Good Home" dogs do not normally pass through Cedars' gates. In fact, "Free to a Good Home" animals had long been the source of research dogs for Cedars and the VA. These institutions bought from USDA licensed dealers whose criminal histories read like rap sheets.

USDA dealers Bud Knudsen and James Hickey were the principal, longtime suppliers to Cedars; Young later defended them, saying none had actually been convicted of pet theft. Those same kennels which eventually lost their licenses had been documented by local law enforcers as filthy, disease-ridden warehouses for pets stolen from streets, yards and "Free" ads.

Only two months before the trial began, in a raid at dawn on January 9, 1991, the Los Angeles SPCA intercepted at Cedars a shipment from Oregon of allegedly stolen dogs. The origin of that transport was a USDA-licensed kennel run by David Stephens, "the man up north" whose calls to Budget signaled rushed shipments of animals. At that time, Stephens was working for James Hickey, who had been accepting stolen pets.

The prior year Dr. Young had been notified of the presence of yet other stolen dogs at Cedars: Twenty stolen Greyhounds purchased from Greg Ludlow. Young was contacted several times by attorneys representing the owners of these "fraudulently obtained" dogs, but he did not respond. "He chose to completely ignore my letters," recalled lawyer Victoria Roberts, who was retained by Northern California Greyhound Rescue, Sighthound Rescue, In Defense of Animals, and numerous Greyhound owners.

When USDA, Cedars, and Ludlow's other clients, among them UC Davis and UCLA, failed to act on these theft charges, California Congresswoman Barbara Boxer demanded an FBI investigation of the dealer. Ludlow voluntarily turned in his license, still proclaiming his innocence; he was defended by Dr. Young.

Young's procurement of animals for the VA was as suspect. In 1990, the VA purchased animals from a USDA dealer named Dick Mertz, who had been acquiring dogs from Barbara Ruggiero. And while the Veterans hospitals were buying dogs of questionable origin, they were also stockpiling animals that had been potentially stolen for other institutions. Since Veterans Administration hospitals are government facilities, they are not subject to USDA inspection; they are supposed to monitor themselves. The VA was parlaying this exemption into a sidelines business, warehousing animals for institutions seeking to avoid USDA regulations.

Even Joan Arnoldi, the Administrator of the Regulatory Enforcement and Animal Care division of USDA, was forced to concede that the stockpiling situation had gotten out of hand. In an internal memo dated August 24, 1990, she told researchers to clean up their act:

> We have recently been made aware that some registered research facilities are doing research on animals or holding animals at Veterans Administration Hospital facilities which do not comply with standards and that they are claiming they have no authority to make the VA Hospital correct identified noncompliance areas. The research facility is responsible for the animals and for compliance with the regulations and standards, regardless of where the animals may be held.

But, true to form, in the years since that memo was written REAC had taken no action against institutions circumventing USDA's regulations at substandard VA hospitals.

Now in court, Susan asked Young to explain the kinds of animals the VA and Cedars preferred.

"Most often mixed breeds, not purebreds. Mixed breeds more closely represent the human population."

All members of the jury laughed, but Young failed to see the grim humor.

"Short hair is preferred for sanitation reasons," he added.

And cats?

"Short hair, mixed breeds preferred. Mongrels are preferred, since they most resemble humans."

The jury again laughed.

At the close of the day, Judge Schacter warned the jury: "Remember, no Dick Tracy."

It was a peculiar, but relevant instruction. Coincident with the LA trial, the theme of the syndicated comic strip series Dick Tracy was pet theft. Max Collins, the creator of Tracy's latest caper, said his interest in the subject was piqued by ads for an article on pet theft, which was published in a national magazine several months prior to the trial. In the "Crimestopper" series, Tracy uncovered a "black market for stolen dogs." These hapless pets were sold, Tracy told his Chief, to "medical research, pet stores, puppy mills, even to gamblers in dog fights." He asked for a "high tech, top priority unit" to bust the dog thieves. The Chief blew his top. "I know you've always had a soft spot for animals, Tracy," he snapped.

"This is about crime," Tracy argued. "I've checked and there are city-wide reports of pets being snatched, and a good number are valuable pedigree animals. It's not a misdemeanor, Chief, its grand larceny! It's an organized ring, a nationwide black market for these animals. . . ."

June 11–12, 1991

Two weeks later, Norm Wegener was in court to hear the defense begin its case. Its first witness was Rick Spero, co-owner of Biosphere. Rick's eleven-year-old daughter, Francine Nicole, sat in the gallery beside Barbara's latest fiancé.

Two years earlier, Wegener had characterized Rick as "an oily type who tries to be smooth." As he took the stand in a gray suit, Rick looked as if he were on trial for corporate tax evasion, not pet theft for sale into medical research.

Prompted by questions from his attorney, Rick explained his duties at Biosphere. He was on the "sales end." His primary duty was "to contact universities, speak with doctors, and solicit sales." This was done on a daily basis October through December.

Rick explained that he was simply following USDA directives in obtaining animals for research. "I spoke to a Dr. Capucci at the USDA. He said I could use the "Free" ads in the papers. I was in constant touch with Dr. Capucci regarding getting the license."

In early October 1987, a USDA inspector visited the Budget Boarding site and, on October 22, Biosphere, the invisible company within Budget Boarding, was granted its Class B USDA license to sell "random-source" dogs and cats to research. The USDA mailed Biosphere a booklet on the Animal Welfare Act and one listing USDA-licensed research facilities. Rick sent out introductory letters to about 150 of these facilities announcing Biosphere's services: "We provide and guarantee animals of sound quality."

"Biosphere's price list was derived from conversations with medical institutions in preparation for the Biosphere license," Rick told the court. "I spoke with purchasing agents and doctors and, from what I was told, our prices reflected 10 to 15 percent lower than the market."

Eli Guana asked his client, "After talking to various medical facilities and to the USDA representatives, did you feel the activities were illegal or criminal?"

Rick firmly answered: "No."

Most of the jury took notes as Rick explained his instructions to Ralf Jacobsen. He said he told Ralf ("the independent contractor") exactly what Dr. Capucci conveyed: "He was not to sign any contracts that he would provide a good home, make no oral promises."

Since Dr. Capucci had not made an appearance, nor had counsel offered an affidavit substantiating Rick's recollections of these conversations, Judge Schacter frequently warned Guana that references to Capucci were not valid as evidence. Guana disregarded this warning once too often.

"The jury is instructed to strike from the record anything regarding Dr. Capucci," Judge Schacter bellowed.

Guana turned his questioning to harassment by animal rights activists, which Rick claimed endangered his life and Barbara's. The defense's theme of animal rights versus medical progress, which was ruled irrelevant by Shacter, was hammered into testimony albeit through the back door.

Susan was looking forward to questioning Rick. She intended to show the jury the faulty logic of his statements, that the basis of his testimony was lies. Her eagerness was evident as she carried a stack of papers to the podium and launched briskly into Rick's acquisition of the Whitman and Greene dogs.

"You got animals for the purpose of selling them for animal research?" Susan asked.

"Yes," Rick replied.

"With regard to Penny Whitman, when you went to her house and took her dogs, it was with the intent to sell the animals for medical research and make a profit?"

"Correct."

"With regard to Bill Greene, when you went to his house and took his dog, it was with the intent to sell animals for medical research and make a profit?"

"Correct."

"In your mind, were these people looking for homes for their animals?"

"Yes, in many cases. Some just wanted to be relieved of their animals so they didn't have to take them to the shelter."

"Regarding Penny Whitman, did she want to be relieved of her dog?"

"I don't know," Rick answered. "We didn't discuss it. She had advertised she was looking for a home. If she didn't find a home, I don't know what she would have done."

Susan's cool delivery did not waver as she asked, "You specifically directed Ralf Jacobsen not to make promises that the animals would go to a good home?"

"Correct."

"Did you hide the fact that Biosphere was selling animals to research?"

"I wouldn't say I hid it," Rick replied smoothly. "Anyone who did not have a need to know, did not know. Once I became interested in the business I would read the newspaper, read books. I came across articles that discussed arson, arson at UC Davis. Judging from what I read and what I heard, although the business was legal, it was controversial."

In the midst of Rick's testimony, Norm Wegener left the courtroom. He could not contain his disgust.

July 29–30, 1991

On his third day of questioning, Rick appeared haggard. He had slept poorly and his eyes were ringed with black creases. His jowls were slack and his thick hair looked greasy. The court schedule appeared also to have worn on Susan. She looked drawn and seemed the more delicate in contrast to the swarthy man on the witness stand.

But there was nothing temperate about Susan's interrogation of the defendant. The Deputy DA's questioning had gained the momentum of an iron pendulum smashing against a concrete wall. She barely waited for Rick's responses to her rapid-fire questions.

"So before you formed Biosphere you knew from Barbara that she had used 'Free' ads to obtain animals? You had in your mind that there would be criminal penalties attached to making false promises, that you could be incurring criminal penalties? Why were all the boarded animals tagged with USDA numbers? Why the apparent secrecy of Comfy Kennels? What did you, in fact, convey to the people who were looking for good homes for their pets? What was Biosphere's intent in acquiring those pets? What was the maximum number of dogs you got on one day with Barbara? Ten? Were you present when she made those phone calls? When were you supposed to get married? June, when?"

Rick's memory appeared to have failed him. He could not remember his wedding date, or exactly what he had told Ralf, or what USDA told him, or what the pet owners said in giving away their pets. He could not remember "if they were 'Free to Good Home' ads or just ads."

Susan showed him photos of Budget Boarding. "Who is standing at the door?" She asked him.

Rick peered at the photo. "I can't tell. It could have been me."

It was clearly Rick Spero. Even his lawyer had to concede the point.

The concrete wall was threaded with fissures and Susan began hammering away at those cracks. "In the context of your obtaining animals for Biosphere, would it have been (appropriate) to say something like 'This dog would make a good pet? Is this dog good with children? Would this dog like to live in a big house?' "

"No," Rick murmured.

"How about, 'Does the dog like to sleep indoors?' "

"No. It would tend to indicate to the present owner that the dog was going to a conventional good home situation."

"Dogs obtained by Biosphere were not going to be living indoors, isn't that correct?"

"Not in the conventional sense of fireplaces."

The jury laughed.

During a recess, Ralf was asked by a member of the press whether he would take the stand.

Ralf had heard the damning testimony by Doug Cameron, the private investigator who had related the substance of their discussion in February 1988. He had seen the look on the jury's faces as, one after another, the victims identified him as the man who appeared to fall in love with their dogs and cats.

Ralf told the press: "Look what they're doing to Rick. They don't believe him. It would be suicide."

Barbara Ruggiero thought otherwise.

During an afternoon recess, Barbara talked with the author in a relaxed manner about her childhood and her love of animals. She said she always wanted to be around "happy animals." Animals were, for her, a refuge from a household riddled with domestic tension.

Barbara's parents, former New Jerseyites, moved to California in the 60s. She and her sister, Jo, eight years her senior, grew up in then semi-rural Gardena and later in suburban Northridge. Their father, Joseph Ruggiero worked as an electrical engineer for Hughes Aircraft, "your typical engineer personality. Distant, unemotional."

Over the course of her parents' thirty-one-year marriage, they showed little affection toward each other, and they often lived in silence for weeks on end.

Barbara described herself as a "real mommy's girl." While Dolores Ruggiero was not an affectionate mother, she was practical. "If you don't set goals for yourself, you won't go anywhere in life," she often told Barbara. In grade school, Barbara decided she wanted to become a veterinarian. She was single-minded in her pursuit, and joined the 4H Club "because it would look good on my resumé." Speaking about the pigeons and rabbits she raised for show and meat, she said, "I had animals that were my pets, that I could say, 'This one I sleep with' and 'This one is my business.' I'm real good at realizing things are for a specific purpose."

But by the time she entered Carlyle College Preparatory, a high school in Encino, becoming a vet had lost its appeal. "People don't usually bring in their animals and say it's healthy. You don't see that. You see animals that have been hit; animals that need surgery. You don't get to really see happy animals. So I tended to veer away from that."

When Barbara graduated in 1980, valedictorian and president of her class, her new goal was to work in the movies as an animal trainer. "It was an occupation, a profession, and you don't deal with sick animals." While making the rounds researching this new career, she was hired for several potato chip commercials as the "fresh face, girl next door."

Barbara recalled, "I liked the illusion. I like things that challenge, and that was definitely challenging—to make something look a certain way, when it was not that way."

She realized she needed training, and set her sights on Moorpark College. The school was reputed world-wide as outstanding for students interested in wild and exotic animal training, zoo management, and behavioral research of animals in the wild. Its seven-day regime often included twenty-four-hour shifts caring for the nine hundred animals in its own zoo, considered California's fifth largest. Of the 1200 applicants who applied along with Barbara, only forty were selected.

William Brisbee, former director of Exotic Animal Management at Moorpark, remembered Barbara Ruggiero—not her academic performance but her interest in "young, naive guys she could lead around by the nose," he recalled.

Typically, first-year students were apprenticed under seniors who were in charge of particular animals. Tom Samuels,* a former lover of Barbara's at Moorpark, recalled, years later, those days. "Barbara was great at playing the game, being friendly, very sweet to people she wanted something from."

Barbara commuted to Moorpark from her mother's house on Chase Street, Northridge; by then her father had left the family. She lived in a tiny, three-room guest house which was more like a tack room, separated from the main house, an unpretentious one-story ranch, by corrals and rabbit pens.

Samuels described Barbara's quarters as "a dump, hardly ever cleaned. All kinds of shit on the floor. Dirty." There was no bathroom, and Barbara used the facilities in the main house where her mother lived.

Despite a grueling academic schedule, Barbara operated a thriving enterprise called "Barbara's Bunny Farm." Her business cards displayed a hand-drawn picture of a cute Easter bunny. The rabbits were sold as pets, for fur, or for meat. She slaughtered them herself.

By her second year at Moorpark, Barbara was prospering. She was performing well academically; she had secured a slot caring for the animal of her choice; and she had two pliable young men over whom she had absolute control. One was a freshman, the other was Ralf Jacobsen.

*Named changed at request of Subject.

The relationship between mother and daughter had, meanwhile, deteriorated. Dolores complained to her friends that she was "fed up with Barbara, sick of her not paying rent," and of the filthy conditions in the guest house. One of Dolores's friends later told police that Dolores neither trusted Barbara nor believed anything she said. "Dolores classified her daughter as a pathological liar," she told homicide detectives.

On December 13, 1984, the body of Dolores Ruggiero was found lying on a highway embankment. An on-site examination by police disclosed large wounds on the right side of the head and ligature marks around the throat. There was evidence that the body had been thrown down the hill.

The news release issued by the Ventura County Homicide Department stated: "The investigation at the scene revealed that the victim was probably murdered, with the cause of death being possible strangulation and/or blunt force to the head."

The house on Chase Street was put up for sale on December 19. With the money from the estate settlement, Barbara bought, for $8000 cash, Budget Boarding's name, inventory, card files, ads, and the right to run the kennel. The murder of Dolores Ruggiero remained unsolved.

July 30, 1991

During the early-morning court session, Dr. John David Young was again called, this time on behalf of the defense. The jury had not taken to Young, and the defense hoped to regain the stature the veterinarian had lost by his clinical indifference and arrogance. Rick's attorney, Eli Guana, a short middle-aged man with dark-frame glasses, asked Young what advice he had offered Biosphere. Young said he had told Rick about the "volatile nature of the animal rights movement" and the trouble Cedars had had with those groups.

"Is there a reason why the type of research in which you are involved is not advertised to the general public?" Guana asked.

Young answered that most members of the public did not realize that animal research was going on in hospitals and that it was something hospitals preferred not to advertise. "Generally, most facilities prefer to do it in a low-key manner. Most prefer to remain anonymous and ambiguous and low-key. They don't want negative publicity from animal rights groups."

On re-direct, Susan asked Young where B dealers obtained animals.

"Most USDA B dealers get animals from random sources. They are not high-production facilities. In 1987 the USDA did not regulate how dealers got animals. Now there are only two sources: people who raise them on their own premises and shelters."

Susan asked, "In 1987, did the USDA permit dealers to get animals from "Free" ads?"

Young replied, "I am not aware of any written law, none in the *Federal Register* or the Animal Welfare Act," adding that he had received appropriate documentation from Biosphere. Young had told Susan in private that Rick Spero had contacted him in March 1988, three months after Biosphere was raided, and asked whether Young wanted more animals. That bit of information was regrettably not admissible in court.

Shortly before ten o'clock, amid stunned murmurings in the gallery, Barbara Ruggiero walked up to the witness stand. She could just as well have been taking a seat at a dinner table. Poised and gracious, she appeared well rested. She wore a linen-blend beige suit, brown tint stockings, and beige pumps. A gold cross hung prominently around her neck. She had dabbed a bit of blush on her cheeks and looked quite pretty.

When Barbara had settled into the witness box she turned and smiled sweetly at the jury. The men and women who would decide her fate could just as easily have been guests invited to her home.

"If it had been me I would not have been so calm," Susan later commented. "I would be scared to death. I would feel ashamed. I'd show that to some extent I knew I was in deep shit."

In a soft, little-girl voice Barbara answered her attorney's questions about her thirteen-year involvement with animals, her education at Moorpark, her movie work supplying rodents to Universal Studios (a point which later could not be confirmed), and her children's shows called "Barbara's Cuddly Creatures," performed for birthdays and special occasions. While her voice was mild, her responses were sharp, prompt, and clear.

Barbara told the Court, "After doing a TV series, I had a lot of rats and mice. I was trying to—I wasn't trying to breed them, but they were breeding. I was wondering if there was another source (to sell them) other than pet stores. I asked Rick and he said that research facilities used animals and he would check it out." Through Rick's conversation with UCLA, Barbara learned that the rats must be "sterile," so she "was out of luck selling rodents to research facilities."

But there was another possibility: mongrel dogs they could sell with a USDA license to hospitals and university labs. She and Rick discussed the use of classified ads as a source of dogs.

What about posting signs that Biosphere sold to research? Watnick asked.

Barbara visibly shuddered. "I did not want to be physically harmed" by animal rights activists. She sounded vulnerable and frightened.

As for her sources of dogs and cats, she acquired them under what she thought were the USDA guidelines: "classified ads, people who abandoned animals at the kennel. Some were purchased from the pounds. I had done business with them in the past. I made arrangements with an officer." She could not recall his name, only that "he was a black man about five foot six. He saved them for me."

Watnick asked, "Were you accurate and truthful in filling out those forms?"

"Yes."

Watnick asked Barbara about Ralf's involvement in acquiring animals.

Barbara explained: "I believe his finances were—he needed to improve his finances. He asked if there was any kennel work or if he could do anything to make money." She said she never gave Ralf instructions about what to say to pet owners. She had no idea what he said, and they never discussed what he said. No, she never used a false name at Bill Greene's and Penny Whitman's. And she accurately recorded the types of dogs and their owners' names on the USDA forms.

Barbara related how Mona Guinney requested she place Flicka elsewhere "because of the boarding costs." She insisted she found a home for Chuck Ransdell's dog, a customer of a local feed store (she could not remember his name), but that the placement had not worked out. She said she found Ammo a few months later "tied to the front door of Budget Boarding, tied to a beam."

Barbara trembled as she described the outrages perpetrated against her. "I received phone calls, I was surrounded by people, animal rights activists picketing, threatening to kill me."

Barbara as victim was ground into the Court record.

"Did you know that getting animals from 'Free' ads might be a crime?" Warnick asked.

"No." Barbara sounded incredulous at the suggestion.

Under cross examination by Rick's attorney, Barbara mentioned the Phillipses' "marital problems. Terry consumed alcohol to an excess." (Terry denied this accusation.) They had "discharged him."

Ralf's attorney Steve Flanagan, a tall broad-chested man in his mid thirties, asked Barbara about the genesis of Biosphere. "The venture with Frederick Spero began with an innocent remark regarding rats and mice?"

"Yes," she softly replied.

• • •

Barbara had done well. She had portrayed herself as an honest, educated, hard-working young woman who followed all the rules. She had offered a young man a job when he needed money, rid her kennels of a quarrelsome, alcoholic caretaker, and innocently stumbled on the animal dealer business because she had too many rats. For all her care and good intentions, she had been victimized by animal rights activists who prevented her from providing services for the public good.

Susan's moment had finally come, and she was ready. She had heard just about enough. Lies compounded by lies, the glib perpetration of lies. The denials and miscalculations that Barbara had the arrogance to bring into court. Barbara had not only deceived the pet owners, she was trying to deceive the Court as well. "It took a lot of balls to get up there and perpetuate what was so obviously a lie," Susan later said.

With every statement Barbara had made, it became that more important to Susan to expose this woman. The Deputy DA had lived with the case for months; and it affected her like no other. It had become an obsession. Barbara Ruggiero had come to represent what Susan despised most in a human being: greed, deceit, and self-righteousness. Barbara appeared to be capable of doing just about anything and walking away without looking back.

As the two women eyed each other for a brief moment before the session began, Susan later recalled that a terrifying thought crossed her mind: maybe Barbara was telling the truth. Maybe it hadn't happened the way the victims told it.

And that's when it kicked in, as it invariably did at some point in a trial. Susan saw through the smiles and the volubility and into the soul of a human being quite unlike herself. "Barbara was simply not stirred by the same feelings as most of us," Susan later said.

She launched immediately into Barbara's record keeping, People's Exhibit #7, in which the USDA tag numbers were recorded for each animal that came into Biosphere's possession. There were evident discrepancies.

"Chuck Ransdell's dog came back two month *after* you placed it?"

"Approximately."

"That would have been September, not November as the record indicates."

"I could have been mistaken."

"About Tag #12, which came from the pound, what do the records show?"

"It says my personal dog—September 2, 1987."

But the date that dog was "acquired" was November 27, 1987. "You had the dog in your possession *before* you acquired it?" Susan asked with some irony.

Barbara did not miss a beat: "It was not available for adoption. He was my personal pet. He wasn't working out."

Tag #14, Mona Guinney's dog. The date "acquired" was October 27, 1987—the very day Biosphere received its USDA license. "Was that the date you put a tag around the dog's neck?"

"Either on its neck or on the cage," Barbara replied.

"You said the dog 'lost its manners' and as of October 27 it was available for medical research?"

"It was available for adoption."

"Available for adoption means medical research?" Susan asked in disbelief.

"Once again, that's one of the options."

Could she point out any tag numbers of animals which were *actually* adopted?

Barbara indicated USDA tag #12, a Dalmation purchased at the pound. It was adopted by someone who lived at the kennel.

"So your caretaker kept the dog at Budget as her personal pet?"

"Yes, then it was stolen by an activist."

Several members of the jury looked incredulous.

Susan rapidly listed more tag numbers—did Barbara recall the source of those animals? USDA Tag #9: David and Betty Jo McCaughtrey's dog Ox, which Barbara had told the Phillipses was her own pet. Tag #4, a Lab mix from someone named "Kelley." Kelley did not have a last name? Tag #111, a Shepherd mix acquired from Ralf Jacobsen.

"Did he supply his own pets?" Susan asked.

Barbara said she did not know. She had listed Ralf as the owner of dogs wearing USDA tags #114, #115, #129, #130, #131, #31, #32, #35, #90, #127, #128.

Barbara had stopped smiling and her voice hardened. "I went beyond the scope and realm that the USDA required," she said icily.

No, she had never told Terry to remove personal IDs from a dog. She did not remember Candy Sheker. No, she did not put "strays congregating at Budget" into (her) inventory.

Susan returned to the subject of Mona Guinney. Had Mona given her permission to sell Flicka?

"She didn't give me that specific permission."

The Phillipses had testified that Flicka was "a nice big friendly dog," but Barbara now said Mona's dog was "not good with children, rough, an outdoor dog."

Barbara contradicted not only the Phillipses' testimony, but that of all the victims, and of Gary Olsen from Animal Reg and private investigator Douglas Cameron.

Could these people all have lied to the Court?

Susan reiterated Mona Guinney's statement that she did not want Flicka to die in a kennel. Susan then asked Barbara: "Was it okay for Flicka to die at Cedars Sinai?"

The women no longer hid their mutual contempt. Barbara, who so carefully chose weak, easily manipulated men, was faced with a female adversary impervious to her dimpled cheeks and little-girl antics. Something surfaced in Barbara's face. Onlookers would recall in that look a virulent, profound hatred, something "evil."

Ralf cowered beside his attorney. Could he have been recalling his own mother's prophecy about Barbara, that "the devil comes in nice faces"?

Susan showed Barbara Exhibits 69 A-D, receipts from the pounds where Barbara "adopted" nine dogs. Susan asked her about box "MR," which on those forms was checked "No." Those owners had specifically forbidden that their former pets be sold to medical research, an option available to owners bringing dogs and cats to LA County pounds.

But like so many pounds across the country, LA County had relinquished its animals without inquiring into the motives of the men and women who came through its doors.

SEVENTEEN

Justice

"She is the personification of evil."
—Judge David Schacter
September 11, 1991

July 31, 1991
San Fernando Courthouse, Los Angeles County
The previous day, Susan had requested that Barbara deliver to the Court Biosphere's financial statements. Her first question of the morning referred to Biosphere's income.

That figure, Barbara said, was $17,500.

Susan clarified for the Court that the amount represented money earned from sales to medical research within about one month, from December 15, 1987, to January 22, 1988. Several members of the jury stared at Barbara. In later cross-examination by her attorney, Barbara would claim her costs to run the kennel during that period exceeded $10,800. Her net was, therefore, "$6675," divided between her and Rick, she explained. That left her, she claimed, at a pay rate of "per hour, $5.50."

Barbara had shown herself to be a chameleon, changing her mannerism, even her tone of voice to reflect the person whom she addressed. With men, she was the little girl or the coy flirt. But she had been unable to find her ground with her first formidable female adversary. This morning Barbara's eyes looked almost bruised, and she appeared subdued.

Like her partner, Rick Spero, Barbara seemed to have memory failure on her second day of testimony. She could no longer recall whether she had told her former caretaker, James Manees, to obtain animals. She could not recall asking Ralf to obtain animals, did not remember any of the victims except Penny Whitman and Bill

Greene—and even then her recollections were dim. She knew only that she had never lied to anyone.

Didn't she recall substituting for two sick cats two others, among them the Phillipses' cat, Charlie, in a shipment to the Veterans Hospital?

No, Barbara said, she did not. She insisted she did not pay Ralf for animals acquired in 1987, and she firmly denied accepting Mona Guinney's checks after December 1987.

But what of the canceled check dated January 1988 and endorsed by her own signature? Susan asked.

"It was a clerical error in accepting the check."

Judge Schacter interrupted. "When were you aware of the error?"

"When she came in the next month," Barbara replied, smiling at the judge.

"Well, that was in 1987. It's now 1991."

"I would be more than happy to reimburse her," she said in a small voice.

Norman Wegener had slipped into the gallery during Barbara's testimony; he was not about to miss this event. As he observed Barbara's maneuvering—"batting her lashes trying to score points with the jury and the judge"—he recalled their first encounter. "She tried to attract me into her web during hearings in municipal court," he later said. "I said to her attorney, 'You can tell your client she's wasting her time.' This woman is an amoral personality. Manipulative, avaricious, cold-blooded. She just never lets up."

Wegener would have loved to cross-examine Barbara Ruggiero, and he envied Susan this day in court.

Now Barbara could not recall Chuck Ransdell's request that she place Ammo in a good home or his stated willingness to take Ammo back.

Susan asked, "When you picked up dogs, you never used fake names?"

"Correct."

"Have you ever used a false name to obtain or sell animals?"

"No."

"Does the name Kimberly Lynn Christianson mean anything to you?"

The defense counsels erupted in a barrage of objections, and Judge Schacter instructed the bailiff to remove the jury.

"Request to cite the District Attorney," Watnick demanded.

The events surrounding Puppy Pavilion, Barbara's most recent pet-theft ring which she ran under the name Kimberly Lynn Christianson, occurred after the events for which she was now on trail, Watnick argued. As such, Puppy Pavilion was inadmissible due to its "overriding prejudicial effect."

Susan countered, "Ms. Ruggiero just told the jury that she never used a fake name."

Judge Schacter was furious, though not at Susan. "You pose the question in the timeframe 1988 or before and I will tell Barbara Ruggiero that she just lied under oath," Schacter boomed. "It's perjury!" But Barbara's later scam was ruled inadmissible in court.

Late into the afternoon, Susan called a series of rebuttal witnesses—a request Judge Schacter had reluctantly granted. It was clear that Schacter had heard his fill. "I'd like to forget some of their names myself," he said of the defendants before the jury entered for yet another session. But since additional victims had been mentioned during the defendants' own testimony, Susan had grounds for her request.

Barbara listened impassively to Betty McCaughtrey, whose dog, Ox, Flicka's cagemate, had been obtained through the "Free" ads; Animal Reg officer Linda Joyce Dean, who said Spero had denied he sold animals to research; and Gayle Sovia, whose dog Barbara had claimed was her own and now refused to relinquish. The last rebuttal witness was Candy Sheker.

Candy had raced to court from her Sylmar home on a last-minute call from Susan. In the hallway awaiting her turn, she spoke about Pooches and her eyes filled with tears. When she retrieved him, he had been very sick with kennel cough, and had a bad skin condition. Once an avid swimmer, Pooches cowered when she merely turned on the garden hose. He refused to swim for a year. "Someone must have sprayed him in that place with a high-power nozzle," she said, her anger still raw.

Ralf had told Doug Cameron that he had found the black Lab on the street and was ordered by Barbara to bring the dog in for tagging. "Maybe Pooches did get out of the yard, or maybe he took him out, but that bitch put him in her inventory," Pooches' owner said. "If that's not a sick brain."

As far as Candy was concerned "the whole thing is a cover-up.

Animal Reg, sure, they knew something was going on. Nothing was done about it. And the research facilities, why weren't they brought up on charges? They were paying $350 a dog, for God's sake! That's our tax money and what do we pay it for—for them to steal our dogs? You got big bucks in those hospitals. Cedars is the hospital to the 'stars.' This makes their place look bad. But that bitch is just one little crook. Who knows where else these research places are getting their dogs and cats?"

The bailiff opened the door to the courtroom and motioned Candy in. "I'm not an animal rights activist," she said, smoothing her tee shirt over skin-tight leggings. "I'm just here because that woman took my dog." Candy strode into court on high-heel sandals, a big, blond, straight-talking woman who was "not about to take any shit from anybody."

Candy told the Court about her search for Pooches, which led her to Comfy Kennel, only twenty yards from her house. There she had met a dark-haired woman and a young man.

"I asked her if she had seen a black Lab with a canvas collar. He had ID on."

Susan asked her if those two individuals were in court.

Candy immediately identified Barbara. She then scrutinized the other defendants and the courtroom. Something was wrong. Slowly Candy shook her head. She did not recognize in court the young man she saw that day at Comfy.

Susan was stunned, but Ralf's lawyer could barely contain his excitement. He motioned for Ralf to stand and then he addressed the Court: "Let the record reflect that this witness cannot identify any of the defendants."

"Wait just one minute." Candy's voice suddenly resounded in the courtroom. "Wait just a minute." Onlookers would recall Candy standing in the witness box as she pointed a long, polished fingernail at Ralf Jacobsen. "There he is."

Steve Flanagan had not felt optimistic about Ralf's chances of acquittal. He had been overheard telling a colleague during a recess, "I really got a winner, here." Flanagan's sense of doom was only exacerbated by Barbara's testimony.

The separation of camps did not bode well for Ralf, who was conveniently and repeatedly characterized as an "independent contractor."

In talking with Ralf prior to Candy's testimony, Flanagan made his personal feelings known. He told Ralf: "One thing I wouldn't have you do is walk my dog. Maybe pick up its shit with your fingernails." When the attorney realized his comments were overheard, he half smiled. "That's a little defense counsel humor. Sick." Then he shrugged, "Hey, you can even put on a defense for Hitler."

With Candy's testimony, Flanagan must have realized the prognosis looked bad for Ralf Jacobsen.

Late in the afternoon of July 31, the defense and the People rested their case.

August 1 and 6, 1991

At 10:12 a.m. on August 1, six months after the trial began, Susan Chasworth stood for her final remarks. During these last days she had taken a nearby hotel room to conserve energy and save commuting time. She looked tired and pale, her eyes bloodshot, but there was a steely reserve from which she would draw for her closing remarks.

Susan had made a commitment to herself early on to try this case on the facts, not on emotion. But as she now addressed the Court, she felt propelled by her own rage, which had crystalized and hardened during Barbara's testimony.

Susan reviewed and explained for the jury the charges: Grand Theft Dog, Petty Theft (cat), False Pretenses, Intent to Defraud, Embezzlement, Conspiracy. She spoke of the strange coincidence that on October 26, 1987, Barbara began stockpiling animals, the very date Biosphere obtained its USDA license.

And from whom did the defendants obtain these animals for sale to medical research? People whose ads and testimony showed an overriding concern for the well-being of their pets. People whose lives were profoundly, in some cases permanently, damaged.

The sense of personal violation would, she said, remain with many of these victims all their lives, as would the memory of the criminal acts committed in their own homes. She reminded the jury that Ralf had kissed John Martin's dog on the muzzle. He had gotten down on his hands and knees to play with Elaine Goodfriend's dog. He had told Jeff Shafter he was taking his dog to Canyon country. He told

Michelle Zelman he wanted her dog for his girlfriend and he told Elizabeth Burgos he wanted adult cats so they wouldn't scratch his furniture. Ralf told Norm Flint about his new home and his pregnant wife.

"As to Greene and Whitman," Susan continued, "Barbara and Rick directly conspired to misrepresent their intentions for the purpose of selling those dogs to medical research. Greene was told by Spero and Ruggiero that they had a ranch in Tuhunja, that they wanted a big friendly dog to be a companion to their daughter and their horse. Miss Whitman they told there was a ranch in Lake View Terrace and they asked specifically 'Is the dog housebroken?'

"There is a variety of misrepresentations and I would submit to you that element one, False Pretenses, has been fulfilled easily by the statements that were made by the defendants in this case."

Susan briefly discussed the embezzlement charges in the cases of Mona Guinney and Chuck Ransdell as a violation of "a type of relationship of trust and confidence. Regardless of what you find as to Jacobsen, it is quite obvious that Spero and Ruggiero conspired together, they were business partners, they prepared a kennel together for business, they got a license together. They solicited their orders from hospitals."

Barbara glared at Susan as the Deputy DA bore down on a point which the defense had attempted to subvert for its own aims. "This case is about theft. It *isn't* about animal research, whether it is a good idea or not. What is important in this case is how these people's property—the property happens to be live things, they're animals—how their property was gotten from them, what was said, what was done. It isn't about what the USDA has said or done. It is about what was said from these people to the victims in this case.

"And that is what you must concentrate on, because those are the operative facts of the offense. If you look at that, at the evidence, if you access the credibility of the witnesses, I think it is real obvious what was going on."

Barbara's face was a dark mask of controlled hate as Susan returned to her seat.

Rick's attorney, taking a folksy approach, opened with an anecdote about a father and son. The story appeared to be lost on most of the jurors, and those who understood it seemed to be unimpressed. Eli Guana then sketched the innocent birth of Biosphere: "Mr. Spero called up UCLA wanting to get rid of a lot of rats. He learns that UCLA and other facilities have use for dogs and cats. Mr. Spero called the USDA and asked: 'What do we have to do?' The USDA told him, 'You have to have a kennel. You have to have a place. You have to have five days.' He went about doing it. Totally legitimate. Totally legal. No written contracts, no oral promises, no misrepresentation. There was a business opportunity they wanted to avail themselves of . . . and there were people who wanted to get rid of their dogs and cats. Of course, researchers wanted mongrels medium-sized. Friendly, you bet. These people are going to be handling these animals. Makes common sense."

Then Guana came to the crux of the defense's position: "I believe that (because of) the events of the animal rights movement, that this case, has become a media event." He asked the jury whether they could possibly believe that his client, Rick, "after having talked to the USDA, after having obtained all the necessary licenses, and after having talked to UCLA, Dr. Young, and all the rest of them, that he says 'Aha, now I'm going to go out there and defraud all pet owners who put ads in the newspaper . . .' (But) if you believe that it was his intent to follow the letter of the law and maybe even come a little close to the edge, because the people wanted to get rid of the dogs, the animals, but he didn't cross over——"

"Your Honor," Susan objected, "that misstates the law."

Her objection was sustained.

Guana closed with another anecdote—this one about a guru and a young bird.

It was midafternoon when Barbara's counsel began his closing remarks. Lewis Watnick paced the courtroom as he returned to a familiar theme: the coercion of animal rights activists over the DA's office, the Department of Animal Reg, and the press.

In a sandpapery voice, Watnick reiterated the defense's opening remarks made on March 26. He told the jury, "This case goes beyond merely the trial of three people charged with an obscure section of the Penal Code, selling animals for medical research. I indicated it was a concerted effort to stop the sale of animals to medical research by the animal rights people and by the prosecutor. What the prosecution is attempting to accomplish is to bring the selling of animals for medical research into the criminal courts, the same as the right-to-life activists are trying to bring abortion back into the criminal courts.

I submit these are moral and ethical considerations, not issues for criminal court."

Watnick, who had derided the DA's office for "making this a case about animal research," proceeded along those very lines. "Who bought these animals? The VA, Mira (*sic*) Loma, Cedars Sinai? Maybe they can be found guilty of conspiracy also. And believe me, if the animal rights people had their option, they would be parties here, too."

He assured the Court that Ruggiero and Spero followed USDA, research facility, and Animal Regulation rules. "They were licensed to sell these animals to medical research. They also received city, county, federal licenses. The USDA has pretty good control in this field," furnishing his client with not only a license but the names of authorized, licensed research facilities and price lists. "Now, can we hold it against them because it was profitable? It appeared to be a good business venture. . . ."

Watnick proceeded to refute the DA's charges of Fraud, Conspiracy, and Embezzlement. "Mr. Greene never asked for last name, phone number, address." In the case of Penny Whitman, the dog was not sold to research for two and a half months. Chuck Ransdell tried to get rid of his animal. Mona Guinney's animal was boarded for over two years, kept in a kennel; so when Barbara had the opportunity she gave it up for animal research. A cage is a cage, after all.

Again, Watnick underscored the animal rights issue over the criminal, steering the case from its legal foundation to an emotional one. "These (animal rights) groups frightened these people to death. But mere pressure is not enough, they had to exert political pressure to be valid, so they exerted their political pressure through the Department of Animal Regulation." It was not his client's fault that "there are unfortunately groups and persons who feel that animals are more important than people."

Steve Flanagan sighed inaudibly as he stood for his final remarks. He looked like a football player past his prime. At this point, his only hope was to embellish on the transfer-of-blame theory from Barbara and Rick to hapless Ralf, from the buyers of animals to their suppliers. Ralf was an innocent pawn, not only of the two dealers but of the system, "the nonindicted co-conspirators: Cedars, Young . . . all those other people who turn a deaf ear and turn their eyes away as to where the animals came from. I mean, there is this little fix that you have to keep the animals for five days, but that is it. They know where these animals are coming from, but they are so quick to dump it on everybody else." The defense rested its case.

In final rebuttal, which is granted to the prosecution, Susan went for the throat of the defense, aggressively disclaiming their arguments as "irrelevant and dangerous." She slammed their contention that the DA's office was the "bad agent, in particular, myself because apparently we've opted to enforce a law that isn't violated that often."

But the pivotal deception created by the defense which Susan wanted to set straight was the "big smoke screen which is a dangerous, reckless and incorrect insinuation by counsel. When counsel argued that the goal of the prosecution was to outlaw the sale of animals to research, there is a clear insinuation here that if these defendants are found guilty that there will be no more sales of animals to medical labs. That is simply not true. This is not the goal of the prosecution— moreover, that is *not* what the result of a guilty verdict in this case will be. The only impact of finding these defendants guilty is that they are the ones who are ultimately responsible for what they did."

Judge Schacter read the deliberation instructions to the jury and the court was recessed until a verdict was reached.

The defense predicted a hung jury.

August 7, 1991
While the LA jury was deliberating, the Italian government made an unprecedented move to stop the kind of crimes committed by Ruggiero, Spero, and Jacobsen. The Italian Parliament had declared "*Basta!*" to the wholesale abandonment of family pets during the August holiday exodus when 60 million Italians left their respective cities each year for the shore. The familiar saying in Italy, "Under the Christmas tree at Christmas, onto the highway in August," had dumped 1.5 million abandoned dogs onto Italian streets within the past ten years. The result of this customary cruelty: 45,000 highway accidents, leaving 1500 motorists injured and eighty dead. Fifteen of every 1000 accidents on the Italian toll roads were caused by stray animals, mostly dogs.

The August 7 issue of the *LA Times* reported that a new law had been approved by Parliament just prior to the summer recess: Anyone caught mistreating or abandoning a dog would face a fine of up to $25,000. Any summer entrepreneur caught rounding up a stray dog for sale to a laboratory that practiced vivisection could be fined up to $8000.

August 9, 1991
At 3:30 p.m. Judge Schacter was handed the brown envelope containing verdicts on fourteen counts of False Pretenses, Embezzlement, Conspiracy to Commit, and Conspiracy to defraud. He perused the verdicts, then handed the envelope to the clerk, who read aloud the findings on each criminal charge:

"Count 1: on or about November 1, 1987, the crime of Grand Theft Dog for Commercial Use in violation of Penal Code 487, a felony, was committed by Barbara Ann Ruggiero, Frederick John Spero, and Ralf Jacobsen who did willfully and unlawfully steal, take, and carry away a dog of another, to wit Bill Greene and Lorraine Greene, for sale, medical research, or commercial use . . . NOT GUILTY."

There was a collective intake of breath in the packed gallery.

"Count 2: On or about November 1, Barbara Ann Ruggiero, Frederick John Spero, and Ralf Jacobsen . . . who did willfully and unlawfully steal, take, and carry away a dog of another, to wit, Penny Whitman, for sale, medical research . . . NO VERDICT." A hung jury.

Susan's face looked pinched as she stared down at her hands clasped tightly on the table; the whites of her knuckles showed. At the defense table, Barbara was beaming. Color had returned to Ralf's face and a hint of a smile hovered on Rick's lips. In the stunned silence of the gallery, Norm Wegener shook his head in disbelief.

"Count 3 . . . in the theft and sale to medical research the dog of another, to wit, John Martin . . . GUILTY.

"Count 4 . . . the dog of another, to wit, Elaine Adler (Goodfriend) . . . GUILTY.

"Count 5 . . . the dog of another, Jeffrey Shafter and Kathryn Malloy . . . GUILTY."

And on and on, a roll call of guilty verdicts on Counts 6, 7, 8, 9,

11—the theft and sale to research of the pets belonging to another, to wit: Michelle Zelman, Elizabeth Burgos, Richard Olsson, Chuck Ransdell, Paul Iverson.

On Count 15, Embezzlement, False Pretenses in the case of Chuck Ransdell's payment of $35 to place Ammo, the jury hung.

"Count 12 . . . the dog of another, to wit, Norman Flint . . . GUILTY.

"Count 10, Barbara Ruggiero on charges of Embezzlement of Mona Guinney . . . GUILTY."

On Count 13, Conspiracy to Commit, 487G of the Penal Code, the most comprehensive of violations which would determine the trio's unequivocal guilt: "On or between October 1, 1987, and January 31, 1988, Barbara Ann Ruggiero, Frederick Spero, and Ralf Jacobsen did willfully and unlawfully conspire together and with another person and persons whose identity is unknown to commit the crime of stealing a dog for the purpose of medical research, a felony . . . GUILTY.

"Count 14: Conspiracy to Defraud another of property. . . GUILTY."

Barbara turned to look at her sister, Jo, who was crying, but her own face reflected little more than annoyance. Ralf seemed to shrink into himself. Rick held his head in his hands.

• • •

In the hallway the press converged noisily on Susan. Was she pleased with the verdict? Was she surprised at Count 1's "Not Guilty"? What about the hung jury on Count 2? What kind of sentence did she predict? Would this set a precedent in trying pet-theft cases?

For the first time since the trial began, Susan Chasworth smiled in public. Yes, she was pleased, very pleased, with the verdict. She could accept the jury's giving way on Count 1 since the conspiracy to commit and to defraud redressed that verdict. (She would later learn the jury hung 10 to 2 for Whitman.)

Yes, she would recommend maximum sentences, the highest terms, and denial of probation. "The scope of the crime, the vulnerability of the victims, the calculated cold-blooded nature of the plan, the heartless greed . . . ."

Susan disengaged from the crowd and returned to her office. It was, she was relieved to find, empty. She slid, exhausted, into her chair, tears gathering behind her eyes.

September 11, 1991

The gallery was packed. Norman Flint had again traveled from Orka Island. Mona Guinney was once again accompanied by her daughter-in-law. Jeannie Plante, who never could learn where big, good-natured Butch had died, had brought her son to court. In the three years that had elapsed since Butch was taken, Montana had grown taller than his mother. Jeannie's eyes were swollen from crying.

Candy Sheker was also present. Outside the courtroom her husband waited with Pooches, who had already drawn considerable attention from the press.

Several friends and family members of the defendants clustered in seats behind the defense table. Each wore a daisy, as a sign of solidarity.

The judge prefaced his remarks with the observation: "Animal rights organizations or their people were not on trial. The use of animals for medical research was not on trial. The theft of animals under false pretenses was the key issue presented to the jury."

The entourage for the defense murmured angrily, and the judge warned against any outbursts.

Schacter continued: "The 'license' defense—that the defendants had a USDA Class B license allowing them to sell animals for medical research—was nothing more than a bootstraping attempt to justify illegal acts." As for the letters on behalf of the defendants that had swamped his office, the judge determined, "These letters tend to show that these defendants led a Jekyll and Hyde existence. . . . Their business dealings revealed their true character. Individually and collectively they were greedy, insensitive, and deceitful."

There was angry stirring in the gallery again on the side of the defense, and the judge put the bailiff on alert. He continued, "Nothing the Court has seen would indicate that these defendants are remorseful for their past conduct or that they intend to abandon their conduct in the future. . . . Miss Ruggiero is a conniving, manipulative individual who has no morals or ethics when it comes to achieving a desired goal. In her dealings with vulnerable pet owners and with animals, she is the personification of evil."

Judge Schacter handed down maximum sentences: *Barbara Ruggiero, 6 years, 2 months; Rick Spero, 5 years; Ralf Jacobsen, 3 years.* The sentences were the harshest ever pronounced in any state or federal court for theft of a pet for sale to medical research.

The sentences are being appealed.

Over a year later, USDA has taken no action on this case except to destroy Biosphere's file, which is no longer available through Freedom of Information.

•   •   •

Several months after the verdict in the Ruggiero case, Mary Warner received a phone call from Captain John McBride. McBride, the astronaut pilot of the *Challenger* (October 1984) had originally contacted Mary when his year-old wolf/Husky, Dakota, was stolen earlier in the year. During that same period, ten big dogs like Dakota had disappeared in the rural area where McBride lived. A "suspicious van" had been repeatedly sighted.

McBride, a member of the Board of the University of West Virginia, got a quick education in the dealer/buncher racket as he scoured the state, accompanied by TV crews and newspaper reporters. His search for Dakota brought him to the Paintsville, Kentucky, dog auction, a popular pit stop in the dealer/research circuit. "There were a lot of good dogs, Mary, and they looked like they didn't belong to any of the people there."

The national scourge had become McBride's personal crusade. "Most people think their dogs wandered off," said the former astronaut, "or that they got hit by a car. But I *know* they're stolen." He and several neighbors obtained the names of local dealers and staked them out. "We know they're stealing and selling and dealing, but once they suspect anyone's getting close they lay off for a while. But it's not only local dealers that are picking dogs off. We believe we're preyed on by every state around us."

McBride had instituted a tattoo program and was working with legislators trying to change the laws. It was still a misdemeanor to steal a dog for sale to medical research in West Virginia.

Had he spoken with the universities?

"Oh, yeah, and we wrote to them all. I wasn't aware that the University of West Virginia was buying from dealers, but they say they're registered dealers. Just about anybody can get a license. That's the problem with the system."

As the three LA defendants languish in jail, tens of thousands of people like Captain John McBride are still searching for their pets, while USDA licensed dealers earn upwards of a million dollars a year selling these animals to medical research.

Meanwhile the official word from the U.S. Department of Agriculture is still: "There is no evidence that stolen dogs are ending up in research."

What You Can Do to Stop Pet Theft

"The road to the laboratory begins in your backyard."
—*KARK-TV, Arkansas*

In November 1991, a breakthrough state Pet Theft Act went into effect in Oregon. The law made pet theft officially a felony and gave residents access to state records which logged dog dealer purchases. In March 1992, Linn County residents, some of whom were armed, conducted a survelliance that led to the arrest of three dog dealers: David Harold Stephens, age thirty, and his wife Traci Lynn, twenty-one, operating as D&T Kennels (USDA license #92B159) and their broker, Brenda Arlene Linville, thirty-three, (USDA license #92B169). David Stephens, who was Barbara Ruggiero's "man up north" contact, had also been purchasing dogs from Linville, who had been combing the "Free to a Good Home" ads for years. Linville had acquired at least 567 dogs and at least 117 cats in a little over a year.

Linn County Deputy Sheriff Dennis Cole raided D&T Kennels and busted the dealers. The county was no stranger to criminal dog dealing. It had been the home for twenty-five years of James Hickey's kennels, which had forfeited its USDA license due to criminal violations. Stephens had also worked for Hickey.

"It's a pretty lucrative business. There's a lot of money involved," said Deputy Sheriff Cole. "The dogs were sold for $200 to $300 each."

As the three Oregonians were arrested on suspicion of first-degree felony theft by deception of a companion animal, pet owners from Oregon and Washington began identifying their dogs which had been sold to the Stephenses' clients: Cedars Sinai Medical Center, the Veterans Hospital at Sepulveda, the Veterans Hospital at Wadsworth, Oregon Health Science University, and the University of Nevada.

Charlie, a two-year-old Border Collie, was one of the survivors. He had been "adopted" by Linville from a Christian Day Care Center where his job had been to play with the children. Charlie was found at Wadsworth, VA, implanted with tubes, undergoing a five-year invasive study of pancreatic disease.

Citizen action also paid off for a victim in Seattle who decided to fight back. Don Johnson prosecuted the University of Washington dog dealer who "adopted" his dog, Sosha, and then sold her for research. In 1990, a Superior Court in Everett, Washington, ordered the dealer to pay Johnson $10,000: $1000 for false statements, $1000 for breach of contract, and $8000 for "severe anxiety and emotional distress."

As Mary Warner explained: "This is a guerilla war and it is up to citizens to take action." State and federal laws do exist, but only citizens can ensure that those laws are enforced. Pets are legally considered property, often times valuable property, and citizens have the right to demand police and court action.

Don't wait to become a victim of pet theft. Act now to protect your pet and those of your friends and neighbors.

Here are a few "golden rules" for protecting your pet:

- *NEVER leave your dog or cat out alone where it can be seen and taken.*

  Never let your pet outside without a leash or supervision, not even in an outdoor kennel; not in your car unattended; not tied on the street while you slip into a store—not even "for just a second."

  Will this cramp your style of pet ownership? If you are like most pet owners who let their animals outside instead of walking them, probably. Will this save your pet's life? Most definitely.

- Do not use the "Free to a Good Home" ads unless you are willing to thoroughly screen prospective owners. This means calling references, including a veterinarian if they already have a pet and their place of business, and visiting their home. Do not forewarn prospective adopters that you will check references. Never give your pet away without first knowing as much as you can about its new adoptive home.

  A better route is to work through a reputable local humane

society that thoroughly checks prospective owners. You may wish to try your local animal shelter as long as it: (1) does not sell to research, and (2) does not put the animal down or has a prescribed waiting period to which it rigidly adheres, and will phone you when that period expires.

• Tattoo your pet but recognize that tattooing will *not* save its life. Unfortunately, tattooing is *not a guarantee* that your pet will not be taken, or that it will not be bought and used by research. As law enforcers like Arkansas Deputy Sheriff Wayne Tyler have learned, dealers will simply scrape off the tattoo or cut off the tattooed ear. Recently, yet another problem has complicated the tattoo process. Many research institutions and animal-use companies like U.S. Surgical, which have a history of buying from criminal dog dealers charged with or convicted of pet theft, are now sponsoring tattoo services in their communities. This brilliant public relations move has lulled residents into a false sense of security and has compromised tattoo services, which are now put in the position of having to defend the source of their sponsorship.

Nonetheless, tattooing can come in handy in the event you must identify your animal at a dog dealer kennel or research facility. Use a reputable national pet registry like National Dog Registry (1-800-NDR-DOGS), but be aware NDR accepts industry sponsorship, or tatoo through your veterinarian.

• Find out what is going on right in your community. Are there dog clubs or pet registries tracking lost pets; any local humane society that is educating residents about pet theft? Call your local dog pound or Department of Animal Control, both phone numbers available through your Town Hall, and find out what kind of job the dog warden is doing. Can your pound legally sell to research? *If you are living near a pound legally allowed to sell to research, you are at high risk of pet theft.* How many days does your pound hold an animal for reclaim or adoption? If it is less than five days, the pound is violating the law. How many animals has your pound sold to research, and to which facilities? Is your pound or local humane shelter making efforts to adopt out animals and encouraging spay/neuter programs? Bring up these issues at town or city council meetings.

Talk to your neighbors. Are their "lost" dogs or cats missing under suspicious circumstances? Have certain breeds disappeared during the same period of time? Watch the "lost" section of your local shopper guide and newspaper. If there is a preponderance of Huskies, Shepherds, Labs, and their cross breeds—premium research dogs—missing, you can be virtually certain something is going on.

• Call people who have advertised through the "Free to a Good Home" ads and warn them of the potential dangers and the need to ask for references which should be checked *before* giving up any pet to a stranger. You will be surprised how grateful people are when offered this information.

• Ask your newspaper editor to preface the "Free to a Good Home" section with a warning of the risks. Los Angeles papers began carrying those warnings during the Barbara Ruggiero trial and have kept up that public service.

• Find out if there are USDA-licensed dealers and research institutions in your area by calling your regional USDA office. Ask for their "List of Animal Welfare Licensed Dealers" and "List of Animal Welfare Licensed Research Facilities," which indicate, by state, dealers and their addresses. Begin a dialogue with inspectors covering your area. By region of the country these numbers are:

> Northeast: (301) 962-7462
> Southeast: (813) 225-7690
> South-Central: (817) 885-6923
> North-Central: (612) 370-2255
> West: (916) 551-1561

A number you will want to make available to your community is the USDA's Hotline for Fraud, Waste & Abuse:
1-800-424-9121

If you are stymied by USDA or your request is not dealt with to your satisfaction, call your congressperson and the

Secretary of Agriculture directly. Members of the House of Representatives can be reached at (202) 225-2121. Members of the Senate can be reached at (202) 224-2121. If you do not know your congressperson's name, call your library or local League of Women Voters. You can reach the Office of the Secretary of Agriculture directly at (202) 720-3631. Remember, these people work for you, the taxpayer.

- You can obtain information on specific dealers by writing to the invaluable Freedom of Information Office asking for the file on a dealer(s) for a particular year(s):

> FOIA
> Room 600, Federal Building
> 6505 Bellcrest Road
> Hyattsville, MD
> 20782

Inform your community of the presence of local dealers, in a *nonconfrontational* manner. For instance, if you write a Letter to the Editor *do not* name specific dealers—unless you want to spend a lot of time and money defending your First Amendment rights. A general information letter about the problems of pet theft will invariably be welcomed. And remember, if you believe a stolen dog is at a dealer kennel, you have the right to request a law enforcement escort onto that property.

- Investigate your local research institutions. Publicly funded research institutions, including hospitals and universities, are *by law* required to have an Animal Care and Use Committee. This committee is supposed to ensure humane care and treatment of the facility's research animals. Each committee is required to include a layman representative from the community. Find out who your representative is, or offer your own services as a community volunteer. Certain states' "Sunshine Laws" require public research institutions to release information about the source of their animals. Write to the public affairs department and make inquiries about how your tax dollars are being spent.

- Spay and neuter your pets. There really are not enough good homes for them all. Contact SPAY U.S.A. (1-800-248-SPAY) for a low-cost clinic near you.

For more information, or to connect with the Pet Theft Citizens Network* call:

1-800-STOLEN-PETS

This free service is offered by Action 81 in association with In Defense of Animals. The Network is not a pet-retrieval service but a conduit enabling front-line fighters, individuals and communities to track dealer activity. Through such local, concerted efforts pets have been recovered, dealers and bunchers arrested, and government and industry held accountable. Be wary of supposed pet-retrieval services, particularly those sponsored by companies or organizations that charge a fee to find your missing pet. Few if any have track records. Most are thinly disguised public relations or fundraising gimmicks.

Take Mary Warner's advice seriously: "Pet theft is extremely prevalent across the country. Please do not leave your pets alone where they can be seen and taken."

* * *

USDA has historically failed to enforce laws that are already in place to stop pet theft. Perhaps the time has come for Congress to make a formal inquiry into a system which has not only ignored public mandate but fostered worsening criminal conditions. Perhaps it is time to take a serious look at the viability of "random-source" dog and cat dealers. As Los Angeles City Attorney Norm Wegener commented, "The very existence of the Class B license, which authorizes the sale of animals from sources that cannot be scrutinized, encourages illicit means of obtaining the animals."

Americans have a long and distinguished history of demanding that government be accountable to the people. Pet racketeering may be the next frontier at which to exercise our constitutional rights of freedom of speech and representation of our interests by government.

*Name subject to change.

GLOSSARY

Animal and Plant Health Inspection Services (APHIS): An agency within the USDA which is responsible for ensuring the health and care of animals and plants and improving agricultural productivity. Among its tasks:

- detecting and monitoring agricultural pests and diseases
- excluding exotic agricultural pests and disease from domestic markets
- providing scientific and technical services
- protecting the welfare of animals
- protecting endangered species

Animal Welfare Act: Legislation passed by Congress in 1966 as a direct result of rampant pet theft for sale into research and abuses at supplier kennels and laboratories. The express purpose of the Act was to protect dog and cat owners from theft of their pets; prevent the use or sale of stolen dogs and cats for the purposes of research; and establish minimum humane standards for the treatment of dogs, cats, and certain other animals by dealers and medical research facilities. While other federal laws, including the Humane Slaughter Act, the Endangered General Species Act, the Marine Mammal Act, and the Wild Horse Act, exist to protect animals, only the Animal Welfare Act provides limited protection for dogs, cats, and other small and domestic animals. The Act required the licensing of animal dealers and made it unlawful for a research facility to purchase animals from unlicensed dealers.

Bunchers: Also known as unlicensed dealers. These middlemen between dealers and the sources of animals obtain dogs and cats from "random sources" and then sell them to dealers. While a dealer may pay a buncher

$3 to $20 per animal, the dealer will sell that same animal to research for upwards of $500.

Class A Dealers: USDA-licensed suppliers of dogs and cats to laboratories. These dealers must breed and raise every animal they sell on their own premises.

Class B Dealers: USDA-licensed suppliers of dogs and cats to laboratories. These dealers are permitted to obtain animals from "random sources." Legally, their sources are pounds and shelters that are authorized by state or county law to sell animals to research; and other dealers who breed and raise the dogs and cats on their own premises.

Office of Inspector General (OIG): An agency within the USDA which performs audits and investigations of programs administered by the Department of Agriculture. Its investigative efforts are directed at improper USDA management actions that result in fraud, waste, and mismanagement.

Office of Management and Budget (OMB): A powerful White House level agency which determines the annual budget of virtually all federal agencies and has the sweeping authority to approve or reject regulations.

Pound Seizure: As required by law in some states, the practice of purchasing dogs and cats from local animal shelters and pounds for the purpose of resale to laboratories. Presently, pound seizure is banned in twelve states; mandated in six states; and left to local jurisdiction in thirty-two states.

Random Source: Sources of animals for use in research other than those specifically bred for research. While legally "random sources" must be confined to authorized pounds and shelters and other dealers who have bred and raised the animals, random sources also illegally include neighborhood streets where dogs and cats may wander or yards where they may be left unattended; auctions where the source of animals often cannot be verified; and the classified ad section in newspapers where pets are offered "Free to Good Homes."

Regulatory Enforcement and Animal Care (REAC): A subdivision of the Animal and Plant Health Inspection Services created to provide leadership and services to achieve compliance with the laws and regulations governing the health and care of animal and plant resources, and to protect the public interest. The Animal Care division of REAC is specifically focused on the Animal Welfare and Horse Protection Acts. Its Regulatory Enforcement division also contracts with other USDA agencies to enforce regulations concerning a range of animal and plant diseases and animal welfare.

United States Department of Agriculture (USDA): The fourth largest federal department, with an annual outlay of about $62 billion, USDA allots funds for various state government programs involving animal and plant health, and food assistance programs including child nutrition and food stamps. In addition, USDA is responsible for enforcing animal protection legislation in the form of the Animal Welfare Act. As such, USDA licenses dog dealers and research facilities and is required by law to enforce standards for the procurement, care, and handling of dogs and cats used in research.

Violation: A violation of the Animal Welfare Act, regulations, or standards that is documented during USDA inspection. Violations include purchasing dogs and cats from unlicensed dealers or from auctions and other unauthorized random sources; falsifying records; and failure to maintain minimum sanitation, housing, general care, and veterinary standards at dealer or research facilities. Penalties for violations include Letters of Warning, Cease and Desist Orders, substantial fines, license suspension or revocation, and imprisonment.

INDEX